The R.A.M.S. Library of Alchemy

Volume 14

Das Aceton
(The Acetone)

**From the writings of
Dr. Christian August Becker**

Also includes the short tract

Philosophia Maturata: Of the Stone of the Philosophers

By St. Dunstan

R.A.M.S. Publishing Company

Das Aceton
(The Acetone)

From the writings of
Dr. Christian August Becker

Philosophia Maturata: Of the Stone of the Philosophers

By St. Dunstan

Produced by

Restorers of Alchemical Manuscripts Society

R.A.M.S. Publishing Company

R.A.M.S. Publishing Company
117 Rutherford Lane
Stuarts Draft VA 24477

First Edition 2007
Second Edition 2015

ISBN-13 **978-1508863670**
ISBN-10 **1508863679**

Image Processing by Philip N. Wheeler

This book is sold for informational purposes only. Neither the publisher nor the editor shall be held accountable for the use or misuse of the information in this book.

Printed in the United States of America

Table of Contents

Dedicated to Hans W. Nintzel,
American Alchemist
and
Founder of the
Restorers of Alchemical Manuscripts Society
(R.A.M.S.)

Disclaimer

Liability: The publisher does not warrant or assume any legal liability or responsibility for the accuracy, completeness, or usefulness of any information, apparatus, product, or process disclosed. The publisher makes no representation as to the accuracy or completeness of the contents of this book and specifically disclaims any implied warranty of merchantability or fitness for a particular purpose. No warranty may be created or extended by written sales materials or sales representatives. You should obtain professional consultation where appropriate. The publisher shall not be liable for any loss of profit or other commercial or personal damages, including but not limited to special, incidental, consequential, or other damages.

Introduction

Philip N. Wheeler

Das Acetone, or The Acetone, is a German Alchemical text written by Dr. Christian August Becker in 1862. It was originally published in 1862, with a second edition in 1867. *Das Acetone* is perhaps the only Alchemical work largely devoted to the study of the Philosophical acetone. The chemical acetone is the organic compound with the formula $(CH_3)_2CO$. It is the simplest ketone; a colorless, volatile, flammable liquid. The chemical acetone is very likely not the same as the Philosophical acetone.

This second R.A.M.S. edition includes a small part of the text from Becker's original Introduction that Hans Nintzel was missing in his copy of the text. A number of transcriptional errors are also now corrected.

This edition includes the short tract, *"Philosophia Maturata: Of the Stone of the Philosophers,"* written by St. Dunstan in the 10th century. St. Dunstan is believed to have lived from about 909 until May 19, 988. This text was restored from a book published by Lancelot Colson, London, 1668, by members of R.A.M.S. and added to the R.A.M.S. Library by Hans W. Nintzel in 1985.

PREFACE

The more recent times which provoked the curiosity in historical sciences by assimilating the past with the present, have also toned down so far as Paracelsus is concerned, but there have been repeated attempts to gain recognition for his work. However, those works were concerned more with his system than with his medications, the reason being that the system as an abstraction of reason may be looked into and criticized by the thoughts of any time period, while the knowledge of the medications, hidden behind the veil of alchemical language, poses very large problems for science and research. Van Helmont has already proved the error of the Paracelsian system, but held the medications therein in high regard.

My studies of magnetism in 1877 led me to Paracelsus, whose thorough medical knowledge of same filled me with admiration. This caused me to further familiarize myself with his work. The darkness of his language made it necessary to look for enlightenment by comparing alchemical tracts and treatises. Then I realized that the virtually untouched *"Feld der Arkane"* had to be the main goal of culture, and the brilliant healings of *Poterius* increased my interest even more. The intrigue of the mysterious was a considerable motivator in my investigations. Test after test was conducted and I was supported tremendously by those two very scientific minded pharmacists, Drs. Grager and Klauer. I was chiefly concerned with the finding of the pain-killing sulphurs of vitriol. *(Sulphur Vitrioli Narcoticurn Paracelsi)* when in 1835 I came to the discovery of *"Ferrum carbonicurn sacharatum"*. I also found the *"Ferrum diaphoreticum Poterii"* which by sublimation of the gold amalgam appeared as a finely separated metallic gold, however, which may be further separated, as may be seen through a microscope, by a simple precipitation of the gold solution with *"EisenVitriol"* or the Vitriol of Iron.

Contrary to popular opinion, it is very effective, even in small doses, and it proved to be especially effective against rheumatism, particularly "Rheumatismus Cordis".

. I pursued this line of research and discovered numerous medications which are not listed in the Pharmocopaea, but which nevertheless are efficacious in practice.

I had hoped for additional information in Weidenfeld's writings *"De Secretis Adeptorum"*, but the main theme, the *"Spiritus Vini Lullianus"* remained a mystery except for some illusory glimpses, and only now, after more than twenty years of renewed studies, did I recognize the idea of the Acetone in the text. This sheds new light on the medication of the Adepts and brightens many of their writings.

Due to the prejudice of the authorities against alchemistry, I probably cannot count on a large participation in my cause, but now and then, there might be a colleague who is secretly interested in this line of research. For this reason and partly because I want to provide a freeing of the obstructions in this domain and because I desire to leave the 70 years of my research as an endowment, I make this small writing known to the general public.

Dr. C. A. Becker
Mulhausen, 30 June 1862

INTRODUCTION

In the old chemical writings, pages 10—14, the characteristics of the *"Spiritus Vini Philosophici"* are completely given, and only the substance from which it is made is kept in a mysterious darkness by making references to red or white wine. For a complete disclosure we turn, therefore, to that part where Weidenfeld, under the title: *"Menstruum Sericonis Riplei"* (on page 329) says the following:

"Sericon or Antimon — both are fictitious names," according to Dean, red lead (lead oxide) is dissolved in distilled vinegar and is evaporated in a water—bath until a consistency of a green gum appears. This acetate salt is distilled from a heavy glass retort, whereby a clear water passes over. As soon as a white vapor appears, a large recipient is connected and well-luted. Then, when a reddish vapor comes over, the heat is increased and subsequently with the stronger fire, red drops issue forth. At this point the fire is decreased and when everything has cooled off, the recipient is taken off and quickly sealed to prevent the escape of the volatile materials distilled over. In the neck of the retort a white, hard sublimate may be found.

The residue in the (bottom) of the retort, is
black as soot. This soot will be strewn onto a stone
plate, and on page 331 it says: [text missing – see
below[1]]

---

[1] NOTE: At this point in the original German text, several
pages are missing. It seems that there are about 2 such pages
missing. Search of various libraries indicates these pages are
missing in several of the extant copies of DAS ACETON.
However, it was felt that this would not seriously harm the
continuity of the textual material since this is merely an
introduction.

However, there also appears to be a page or two missing from
the very first chapter. This is determined by the fact that
the original German text has what appears to be the first
chapter, starting with a portion of a Latin quotation. Since
it is impossible to determine what those pages contained, no
guess as to how serious this loss is can be made. In any
event, since there have been no other tracts, to my knowledge,
on Acetone, these pages should be welcome to the alchemical
researcher and experimenter.

HWN

According to Hans Nintzel, this short Latin passage was originally at the start of Chapter 1:

Accende in extremitatum altera carbone vivo, et spatio elimidiae horae transcurret ignis per omnes faeces, quas calcinabit in colorum citrinum gloriosum valde.[2]

---

[2] Although Hans could not find the text of the missing pages, this is from another source, and appears to be the start of one of the missing pages. A possible translation of the Latin: "Ignite the coals in the extremities, and for long hours maintain the fire, building its hot coals, by which calcine the matter to the color of yellow, very glorious." pnw

PREFACE TO THE SECOND EDITION

When in 1862 I published this little writing, I believed to have made testament with that, considering my advanced age. However, with the Grace of God, I have lived for a longer time and after a serious case of hemiplegie in April of last year, I have recovered sufficiently to resume my spiritual activities and also continue with my research and studies. Therefore, I consider it practical to give this Introduction and at the age of 75, my fondest hope is to be able to see active and enterprising works in this forbidden and forgotten field. This then would be of great help to uncover the long-hidden treasure.

Dr. C. A. Becker
Mulhausen, February 1867

Chapter I

Das Aceton

This is the place in the whole book where the behavior of the residue is described this clearly, and after having been lost in dark words for years, all of a sudden I was enlightened. The characteristic to burn like a tinder, made it clear beyond doubt that the coal—like residue has to result from the destruction of an acetate salt. Thus the secret of the *Spiritus Vini philosophici* was discovered and all the products from the distillation were correct. Now the *Aqua ardens* with the quintessence became a simple chemical fact, and the only thing surprising that was left was how the old chemists had been able to work with it for centuries without this becoming known. Of course everybody put a curse on whoever would give away the secret, and this curse seemingly represented a moralistic power because Weidenfeld in a book that was to be published later on indicates hope for the discovery, but the book never came out. And Pott, who had thorough knowledge and who did not have to fear the curse says, that whether because of a promise or because of envy: the preparation is easy, but it is a secret.

The tindered yellow residue is dissolved with vinegar, evaporated to rubber and distilled. The residue is again treated with vinegar and also distilled. The distillates are poured together, combined with the previous one, left 14 days to digest, and then distilled. At first the *Spiritus ardens* passes over, which is then rectified until it is so strong that a. linen cloth soaked in it and ignited will burn. During these rectifications a white oil appears on the surface and a yellow oil also remains, which is being distilled with a stronger fire.

The sublimate in the neck of the retort is being pulverized and is placed to melt on an iron plate in a cold place. The liquid is filtered and a little *Aqua ardens* is added, whereby a green oil will separate and settle at the surface. Then it is distilled. First water and then a thick oil appear. The water is distilled in another recipient and evaporated in the water bath until a thick, oily substance, like melted tar, remains on the floor. This black liquid substance is treated further with *Aqua ardens,* however, which is not explained any further here.

The Gold Tincture

Just as gold is considered the finest metal, the hermetics also thought of it as the finest medication, and that is how the *Aurum potabile* was put on a throne upon which it remained for many centuries. But as much as they revered it, their secret dissolution against was honored as much and maybe more, and they called it gold too. In his old age Raimund, for his strength, prepared the raw oil from the lead and said that it was more delicious than gold. Basilius Valentinus, who describes the preparation of the *Spiritus Vini philosophici* under the cover of the distillation of the vitriol, called the raw oil "heavy as gold, as thick as blood, burning and fiery, the real liquid gold of the philosophers."

The ideal of the alchemists and the masterpiece of the art was the *Lapis Philosophorum,* the stone of the wise ones. For its preparation the most needed metal was the gold. The customary gold was not suited for this purpose because it was dead due to the firm closure of its particles, and it therefore had to be animated first. This was achieved by treatment with the *Spiritus Vini philosophici,* whereby the soul, the characteristics were separated from the impure body and dissolved. This gave then the philosophic gold, *Aurum nostrum,* the quintessence, the radical dissolution without corrosion, which was achieved through the raw oil of the *Acetone,* called the *Acetone aerrimum* and the *Dissaeveus Auri.*

This dissolving power is confirmed by experiment reported by Fuchs (Geschichte des Zinks pag. 200). Hellot distilled acetic zinc. At first a lightly acetic phlegm transformed; then stripes appeared, and then followed a sublimate in white, fragile flowers. Then white vapors arose which were condensed in the top of the flask into whitish-yellow, then dark green oil. The recipient contained a liquid which ignited just like *Spiritus vini.* Poured onto water, it first swam at the top, then mixed with the water, and only a few drops of a reddish, spicy oil remained at the top. The residue of the distillation was of the color of ashes. On it the acetic phlegm was poured, digested for 8 – 10 days, then drained and distilled, leaving a resinous substance at the bottom. The process was repeated until enough resin was obtained. This in turn was then distilled in a small retort and heated to the point of glowing, whereby a yellow liquid transformed, followed by thick white vapors. When the distillate was poured on the white sublimate in the neck of the retort, it dissolved the sublimate immediately, and some drops of reddish oil separated on the surface. This oil was then rubbed onto gold and silver trays, which within 4 hours dissolved at the covered areas.

Alchemy through shady lab assistants, crooks, and dreamers has gained such a bad reputation over the years, that in general it is considered to be superstition, cheating, or fraud. Only in more recent times have individual voices of the educated world turned to the expression of Marsilius Ficinus, saying that the old and new philosophers, as the natural scientists called themselves then, have spent much effort and work in order, to explore nature, and they have subsequently recognized the honorable effort of the old chemists. It was natural science in its old form.

Three Essentials

As a basic rule it was established that all bodies are composed of the 3 chemical elements: salt, sulphur, and mercury. The names which are meant only as symbols, and which mean something quite different, would equate with the following in today's terminology:

mercury   =   hydrogen

sulphur   =   carbon

salt   =   oxygen

Missing is nitrogen whose existence as a simple element, however, is still doubtful.

The theory said that the differences of the metals is based on the qualitative proportions of these 3 elements, and that through changes in these proportions, it is therefore possible to alter the metals up to a perfection of gold and silver. Since the proportions of the mixture were determined only hypothetically, the experiment was only empirically technical; but since all metals and many minerals had been used in the experiments, it presented an opportunity for many chemical discoveries which served the sciences in general. The old experts and highly respected persons, like Albertus Magnus and

Roger Bacon, analogously to the efforts of today's science, tended to dismantle the bodies and to create new connections. The old chemistry, using the transmutation of the metals, arrived at no confirmed positive results, whereas modern chemistry not only calculated the atoms, but also their aberrations.

The chemistry, which was taught freely in Arabic sciences and which was protected by the caliphs, encountered mistrust and suspicion with its transition to the Christian world. It was derived from the non—believers whose actions were connected with the world of magic and the devil; it was persecuted by the Church. Working with it was, therefore, socially dangerous, and physically the vapors of the minerals and the vast efforts were not advantageous for the health. Large incentives were needed to find followers and disciples, but they were not scarce. Just as the Church promises its believers eternal happiness, alchemy promised retention of health by means of the *lapis* and with that a longer life and large richness or heaven on earth; in addition there was the secret with its mysterious appeal. Permeated by the grandeur of their ideal, the alchemists drowned themselves in religious mysticism; everything started with God and everything was done under his protection, and only through God's grace and enlightenment could the *stone of wisdom* be obtained.

Aurum Potabile

The radical dissolution of the gold, which was caused without corrosives and from which the metal could not be reduced, was the true *Aurum potabile,* the quintessence. Rupescissa says, "the quintessence of the gold is *Aurum Dei* and part of the *lapis,* and it is completely transformed into *nutriment.*" The genuine gold is not transformed into *nutriment,* but it is excreted in the form in which it is taken in. *Aurum alchymicum,* which is composed of corrosives, destroys nature; therefore the *Aurum Lapidis* is called *Aurum Dei.*

Paracelsus explains that the quintessence in the gold is very little, but that it has the power in the color, and when it is extracted the remaining metal has lost its power. It differs from *Aurum potabile* in as much as it may not be reduced to metallic gold a second time, while the *Aurum potahile* may be transformed into a metallic body; therefore its quintessence is finer.

Raimund gives the following eloquent, but complicated statement:

1.   *Spiritus Vini philosophici* is distilled three
times over *Sal Tartan,* and this distillation is kept
in digestion for 50 days, at the end of which a
yellow residue appears on the bottom.

2.   The gold and silver are now separately
calcined, that is amalgamated and the quicksilver
evaporated.

3.   On the remaining calcinated metal each
separately we pour three fingers' width of the sharp
spirit No. 1, and then first hold in a water bath,
and then in the ash-bath at boiling temperatures.
The dissolution of the gold is yellow and it is
carefully decanted; also the dissolution of the
silver is green or blue and it is carefully
decanted.

4.   The residue of the metals is repeatedly treated
in the same manner until everything is dissolved.

5.   These solutions are each kept 40 days in
digestion; then the solvent is distilled—out in a
water bath, leaving the metals behind just like
oils. The distillate is poured back over the oil,
left to digest in the water bath for 24 hours and
then distilled.

6.   The distillate is at first gently distilled in a sand bath, whereby the water goes over, then at higher temperatures the spirit goes over, and even at higher temperatures a part of the oil transforms.

7.   The water that went over at first in the water bath is added to the distillate, digested, distilled in the sand—bath, and this is repeated as often as needed to have all the gold and silver go over.

8.   The solutions are rectified in the sand—bath 7 times.

9.   Now both are mixed and circulated for 60 days.

With this the great solvent is prepared, which dissolves all metals radically.

Now other gold, which has been amalgamated and calcined through the evaporation of the quicksilver, is digested with the solvent No. 1, and after its distillation it is submersed by *Menstruwn majus* to dissolve the gold. When this is done it is drained. On the residue a fresh *Menstruum majus* is poured for complete dissolution, and that is then combined with the previous one. The solution has the color of a most beautiful ruby or carbuncle. It is circulated for 20 days in a water bath and 20 days in an ash-bath. Then you will find the gold transformed at the bottom into beautiful rosin, and the water on top may be carefully drained. The rosin is soluble in any liquid. This is the true *Aurum potabile.*

The procedure is described so clearly that with the exception of the secret solvent agent it is totally understandable. Noteworthy is that not only gold, but also silver is needed.

Rupescissa's procedure is simpler. Gold amalgam is finely atomized by evaporating the quicksilver, and then after adding *Acetum philosophorum* it is placed in the sun. This causes an oil film on the surface which is taken off as it is forming, and which is placed into a glass with water. The water is evaporated and the quintessence of the gold, which contains the highest sweetness, remains.

A similar case should be the *Essentia dulcis* of the orphanage in Halle. According to the report by Dr. Richter, its inventor, the essential element is a subtle red gold, which dissolves quickly and without residue or turpidity in the "spirit of wine." When the alcohol (spirit) is taken out, a blackish powder remains which may easily be transformed into a light, fragile, purple red and sweet powder: there is a slight weight loss during this process because the most subtle, even at low temperatures, rises in the form of vapor, which when caught, condenses into red drops.

The way to prepare the gold is very different from the usual method, and even though harmless minerals are needed for the preparation, all foreign additives are so separated that all samples can prove that no corrosives are contained therein.

Half an ounce of the ordinary essence costs 2 "Taler."

Half an ounce of the concentrated essence costs 8 "Taler," since the latter contains 4 times as much gold. The substance was considered to be too. expensive and it was said that the gold part hardly amounted to 1/8th of the price. The gold is, however, the least, but the other expenses and the efforts for the preparation, which keep several people busy year after year, are such that the price in comparison with prices of other medications should be set higher.

In Crell's records of 1747 the doctor of the orphanage, Dr. Richter, a grandson of the inventor, states that the process will be revealed in time. I cannot find any news in regard to this and an explanation is to be expected from Halle.

This information is very little, and therefore a more eloquent report by Woliner *(Diss. inang. de Epilepsia ejusque medicamento specifico Essentia dulcis adpellato. Lugduni Batavorum 1706, 4.p.22)* should be mentioned. According to him it is prepared from purest gold, which is so refined that even the simple *Spiritus vini rectificatissimus* will dissolve a large amount of it and then turn ruby-red. The characteristics attributed by chemists to the radically dissolved *Auro potahile* are found also in the *Essentia dulcis,* that is, it cannot for the most part be reduced to a metallic body, but it evaporates like smoke even with medium fire. When a large enough amount of water is poured onto this essence, it turns turbid at first and then a very fine powder sinks to the bottom, which when dried in mild warmth, shows a yellow color and a bitter taste. It is however of such finesse that when added to spirit of wine, it dissolves completely like wax and it represents again the *Essentia dulcis* in color and taste. This indicates that the color of the *Essentia dulcis* originates from this powder or the finest *Crocus Auri.* When this powder is heated a medium temperature in a glass over coals, there will every once in a while be very fine coins of reduced

gold, but the largest part. of the residue seems so
dissolved, refined and freed from all metallic
chains, that it may not be reduced to metal because
as the powder feels the fire the larger part flies
away in smoke, leaving a fine powder which may not
be reduced either with *"Spiessglanz"* (antimony glass
stibnite) or with lead, but which forms a highly red
purple colored salt when melted with *Sal Tartari.*
This salt will even perpetrate the Tiegel and color
it purple on its outside.

In 1723 Kleinfelder in Konigsberg issued a
statement against this essence, saying that it was
nothing else but a tincture of burned sugar, and he
said that the sugar tincture that he invented was as
effective as the *Essentia dulcis,* even if it really
contained gold. Later it was believed that the
black, coaly residue of the preparation, when
lengthened with ether to become a reddish-brown
tincture, and when mixed with Franzbranntwein, was
the *Essentia dulcis.*

It seems that the procedure was done according
to Lullius: an indication for that is the
preparation from the black residue in the distilling
of the ether. The wrong interpretation may have
resulted from the fact that Lullius calls the
substance for the *Spiritus Vini philosophiei* in many

places *Nigrum nigrius,* and after the distillation of the acetic salts a black substance like melted pitch remains. Maybe a spy in the laboratory overheard something about this black residue in the retort, and thought to have discovered the "wine spirit-coal" in it.

The earlier hermetics used their *acetone* in many ways, partially for chemical procedures in connection with acids and salts, partially for the preparation of medications. From the vegetable substances the quintessence is extracted within 3 hours when it (acetone) is used. An interesting observation of Rupescissa is that the *Laxantia* through this become more effective and are therefore administered in smaller dosages.

Among the later chemiatrics Quercitanus used it for the preparation of the *Antipyreton* and a gold tincture, and Agricola too manufactured several medications with it without realizing that he already possessed the *Menstruum Lullii,* which he desired so much to obtain.

THE HEALING FROM PODAGRA

Which the Count *Onuphrio de Marsciano* tells of in his hermetic writings of 1774, p. 30.

When he suffered a severe case of Podagra he placed the spirit on the swollen and extremely painful foot, and "Oh Wonder!," he says, the pain disappeared and I started to dance for joy, to the astonishment of my friend. After that the Podagra has not pained me again, and I didn't have the least bit of complaint thereafter, but I have been completely free and healthy like before; but from then on, I started to take 20 drops in the morning before eating for 15 days in order to completely clean the blood since there is no blood cleansing like it in the whole world. He calls the substance only *spiritus simplex,* but in the hermetic experiment on page 161, where he cites from Lullius that the quintessence heals all tiredness and sickness, and removes all weakness, protects from all sicknesses and retains the youth, he clearly says: And I swear the truth that I have seen wonderful things done by this *Sirnplici Spiritu Vini philosophici,* and I have even healed the Podagra completely with it, as many have seen and have been ashamed by.

The newer chemistry has again taken up the research since Chenevix found the acetone as *Spiritus pyro-aceticus;* this research, however, dealt only with analytic interests, disregarding its medical costs; and medicine was left empty-handed without its due share.

The alchemists rectified the pure acetone repeatedly in order to eliminate the water, and to arrive at a concentration that would burn like alcohol. The more modern chemists dehydrate the acetone with calcium chloride, which we however cannot approve of, since the latter combines itself with the wood alcohol which is analogous to the acetone. This combination does not dissolve at 100° this condition proves disadvantageous when the product is used as a medication. This process seems also unnecessary since the *Aqua ardens* (das Aceton) is more volatile than the spirit of wine, and it merges already at 48° in veins, while the water follows only at higher temperatures, and the two oils do so at an even higher temperature.

The entire distillate was kept in digestion for several weeks in the warmth of horse dung (30°), whereby especially the oil the quintessence, is separated on the surface and it provided a very pleasant smell. This oil consists of two oily substances: one, a distillate according to Fittig *(About Acetone* 1858, page 48) at 90°; the other one, *Dumasin* at 120°. These two oils form the core of the medication; therefore the substance is an acetonium oleosum and correctly should be called an Acetonol.

ACETONOL

The pure acetone, as provided by the chemical industry, is of little medical value. It is clear and light as water, burns completely, but has no trace of oil on the surface. The oil, however, is still inside because if you place the acetone in mild digestion over an extended period of time, the oil appears and surfaces. I already observed this reaction in the past, and I have repeated the experiment now. I placed ½ ounce pure acetone in a glass that was not tightly closed on the back burner. After approximately one—half had evaporated, a trace of oil appeared, and after two months, when only ½ Drachme remained, a visible layer of a clear oil was on the surface.

The pure *acetone* may be quite good as a chemical preparation, but therapeutically it constitutes a weakened oil-poor product, which only has the appearance, like vanilla beans, out of which the aromatic Benzo—resin has been drained.

For medical application purposes, it will be advisable and required that it is prepared with the same method used by the hermetics. It takes a lot of time and patience and under the current situation of the business, these may not be expected because already in 1668 during a discussion of the Weidenfeld *Spiritus Vini Lulliani,* Jungken complained that the modern chemists are not able to produce anything extraordinary since they start to work in the morning, but stop again at night, which is the wrong way, because a good thing takes time.

Chapter II

The Wine Spirit of the Adepts

This investigation is based on the work of Johannes Seger Weidenfeld — de Secretis Adeptorum sive de usu Spiritus vive Lulliani Libri IV. 1685. 12.

In the dedication to Robert Boyle, Weidenfeld speaks about the progress of his studies. He had diligently studied the work of Paracelsus 10 years ago, but after two years of study he had gained no clear insights. Especially the unfortunate preconception of the alcahest posed a big problem. Already without hope of being able to learn its preparation, he consequently compared the descriptions of Circulatum minus, Specificum corrosivum, etc., in order to find the method of preparation, while being convinced that all of them were one and the same dissolvent. Numerous and hardly believable experiments proved futile, and he had already planned to give up chemistry and medicine, when his eyes were unexpectedly opened and he realized that they did not only have different names, but that they were different in material, preparation, and use. For instance, instead of the single Liquor Alcahest he found several solvents,

their preparation, and their usage. What remained
incomprehensible to others in Paracelsus, became
clear to him and he reached the end before the
beginning. His joy, however, did not last long
because several futile experiments taught him that
the solvents of Paracelsus contained something else
secretive which could not be taken literally.

With that he dropped the alcahest experiments
and turned to studying Lullius, Basilius, etc. There
he realized that they all agreed and confirmed the
Paracelsus solvents, that the preparation of such
was simple and to be understood literally, and that
only one word remained unknown which, however,
according to the experts identified the general
basis of all such solvents, that is the Spiritus
Vini philosophici, with whose knowledge and
possession the greatest secrets in chemistry were
solved.

In Wilna he heard of Robert Boyle, who was the
only and the first person in chemistry to use an
open and clear language. Therefore he went to see
him in England to discuss the solvents and the
medications of Paracelsus, as well as to discuss
other secrets. Boyle accepted him well, praised his
studies, and therefore increased his ambition for
higher achievements.

It is noteworthy that this Spiritus Vini philosophici, whose composition has been clearly given by Weidenfeld, is not mentioned by the later chemists. Only Pott (Exerc. chym. Berolini 1738. 4.) p. 21 describes it with the following words: There is an oily menstruum which has not been named yet, and which has not been revealed by any chemist. It is a pure, light—colored, volatile liquid like the wine spirit; it is oily and burns with a bright flame; it tastes sour like strong vinegar. During distillation it transforms like snowflakes; it affects all metals and gold, extracting the latter in a red form, and when the menstruum is taken off, the tincture that remains resembles resin which dissolves dark—red in Spiritus Vini and which leaves a black residue from which, as I believe, the Sal Auri may be made. This menstruum mixes with water and oils, and if you ask me for my opinion then I would say it is the true menstruum of Wiedenfeld, the Spiritus Vini philosophici. The preparation is easy and simple, but a secret: and Pott does not reveal it. Wiedenfeld promised an explanation in the 5th book, but this 5th book was never published. Others have prepared the substance and used it as medication, but have not known its identity with the Spiritus Vini Lulliani. The newer chemistry concerned itself repeatedly with it and researched its nature, but it hasn't found an opportunity to connect its research to the works of the experts and to make it available for medical application.

This lets us automatically take a look at the pharmaceutical chemistry. It was the traditional task of doctors to produce and improve their weapons, especially the chemical medications. With the big triumvirate Stahl, Boerhaave and Hoffmarin, pharmaceutical chemistry reached its peak; the arsenal was well equipped. The pharmacists lent a helping hand to the doctors, and since the doctors could rely on them, slowly more work was left to them, and only in single instances would doctors work with research and preparation of chemical medications. The upswing in botanics by Lirine, in the pathological anatomy by Morgagni, in physiology by Haller, and in chemistry by Lavoisier, led the doctors into other fields which promised rich harvests on little worked grounds. The pharmacy followed the immense progess in chemistry quickly, arid achieved an importance which was favored and supported by the government and the doctors. The rights of pharmacists were generously outlined and the secured lifetime position increased the performance and the scientific eagerness of this class. The technical chemistry developed further, however, and led to the installation of chemical factories, thus changing the whole situation. The advantageous position of the pharmacists and the easy procurement of the preparations from the factories, where the time— and money—consuming lab work could largely be saved, caused cheating with the pharmacies. The prices tripled and quadrupled. A pharmacy that was worth 20,000 Thlr. guaranteed to its owner a good income; the new buyer paid 40,000 Thlr., and now the interest on the additional 20,000 Thlr. had to be worked out too. That caused consistent complaints in regard to insufficient taxes, interference with rights, and shortage of protection.

As much as the government tried to help by increasing the taxes of medication and work, the complaints continued because with increasing income the prices of the pharmacies rose and with that the interest for the anti—natural additional capital.

The Prussian government tried to limit the growing power of the pharmacist—order in 1810 by issuing concessions to a person for the newly built pharmacies. In the newly acquired French provinces, all privileges had already been lifted and there were only concessions. With time, the difference between a concession and a privilege disappeared, and the government by "highest order" of 1842 reacquired the free disposition of the government over the concessions. After that it was determined: 1) the concessionaire is obliged to take over the supplies, etc., according to their tax value, i.e., according to their real worth; 2) a competition was to be held and the government reserved the right to give the concession to the most qualified pharmacist.

This was ideal. The concessions herewith became civil service jobs and it was in the government's hands to unite the most talented and ambitious pharmacists in a brilliant scientific chain-work, just as it was doing with the other state employees.

The means for the execution, however, was never used; it consisted in setting the price of the concessions according to the appraised value and to adjust the tax accordingly, so that the amount would be sufficient, but not exorbitant. The pharmacists already holding concessions argued vividly against this supposed limitation of their rights of property, and already after four years the government gave in, put the ideal aside, and went back to the old, narrow ways.

Since then things have continued to deteriorate. The pharmacies have sunken to mere

industrial installations. As industrial enterprises
they also carry the risks of such an installation,
and the government has no obligation toward the
country to support and to help that stock market
game. The need of the time has reestablished the
power of the government. The feudal rights in
Austria, the real rights of businesses in Bavaria
were abolished without damages, and in Prussia too,
the tax freedom of the. "Knight estates" was
eliminated through an appropriate release payment.

In one example I was able to observe the way in
which the concessions were handled. A speculating
pharmacy helper applied to establish a pharmacy in a
village, but was denied permission. He told me
openly that it was not his intention to keep the
pharmacy, but that after a few years during which he
would create a booming business, he would have sold
it, expecting a profit of 6,000 Thlr., with which he
would have been able to start something new.

In this situation of the pharmacies where the
lab has lost its old, honored significance, the
doctors are earnestly reminded to concern themselves
again with the preparation of chemical medications.
Considering the enthusiasm for chemistry, a number
of capable and doubtlessly big results could be
gained. That will also serve a good purpose, that

is, that through the self-involvement a large trust in the medications is being achieved. The complaints concerning the unreliability of the medications and the deficiencies of the therapy will disappear because they are due to the fact that most of the young doctors lack the practical knowledge of the medications. They don't know their weapons, and therefore don't know how to use them.

Surgery has a large assortment of instruments available. No surgeon has been before an anvil; but the steelworker manufactured all instruments, but none of them has invented any. But the surgeon in his mind invented the instrument according to his needs and the steelworker only executes the surgeon's idea. Just like the surgeon who cannot be without the instrument maker, the doctors cannot be without the pharmacist; but both are only helpers, not leaders. The wrong approach, where the pharmacist pushed himself into a leader's position, has brought great damage to practical medicine. Many of our best medications originated in the old days, and their application today is based upon the recommendation and 'previous observation. The intended supposed improvements of the formulas are frequently nothing but falsifications. Another mis-take was the change in names of the medications and the adaptation of the respectively reigning chemical

theory. Hufeland requested in the name of the
practical doctors that the old names be kept, but
the governing pharmacy found that to be below its
scientific honor, and only with special
consideration did they compromise to add it in
parenthesis. Mercurius dulcis and Calomel are old
names for a common medication back then, and the
doctors held onto the name by exception, but the
"Pharmakopoen" list it under more than seven names,
all of which were scientific or which had been
partially changed back to "unscientific" ones. The
old ammonia lost its real name through the chemical
baptists. This way leads to the Tower of Babel, and
if the doctors keep changing to old ways, soon they
will not be able to communicate anymore with the
pharmacists.

HELIAS ARTISTA

Paracelsus repeatedly expressed a prophecy which his followers accepted truthfully, and which merits remembering in the interests of history. The references are:

1)   From the preface to *Tinctura Physiourum,* German version, Part I, p. 921.

My theory, which is based on the enlightenment of nature, may not be reversed in its consistency, and it will start to bloom in the year 58. And consequently the practice with unbelievable signs and miracles will prove that also the workmen and all the common people will understand how the Theophrasti Art stands up to the muddling of the sophists who because of their incapability want the protection and support through papal and imperial liberties.

And on p. 924.

These "arkanes" which cause the transformations are little known. And even if they have been enlightened by a god, no immediate glamor of the art will appear, but the Almighty also provides them

with the reason to keep them secret until a future
time, *Helias Artista,* when the secret will be
lifted.

2)   *De mineralibus.* Part II, page 133.

It is true that the earth still holds much of which
I know nothing; others, too, have no knowledge. I am
certain that God will still show many strange things
which never have been shown and of which we did not
know anything. It is also true that nothing is
hidden that will not be made known; therefore, after
me there will be someone whose magnificence is not
yet born, and he will make it known.

3)   Of the natural things, Chapter viii. About
Vitriol, Part I, p. 1506.

Therefore I say that many secrets lie in
nature, in other things of nature, and in God's
creations, and would be better and more useful to
study such things rather than indulging in drinking,
whoring, and other mischief. But nowadays the
whoring will be going on until one third of the
world has been murdered; the other third dies
because of roguery, and hardly one third remains.
Then things fall back into their places. But as
things are going now that might not happen. It is

also necessary to extinguish the caste system in the world or it might not happen either. Then we will have the golden world; that means then man will come to his senses, live like a human being, not like an animal, not like a pig, and not in the "dives."

When will that be?

While some have anxiously awaited the *Helias Artista,* others do not look at it as a person, but as an expression of a period when science will be at its best and will be a common asset to all.

That time has started with the new chemistry, and if you look at our time during this period, and if you want to personify the *Helias Artista,* there will be no doubt at whom in Germany all the educated people will be looking.

The theological opinion moves the goal even further. Hapelias, who in Vol. VI, *Theatra Hemiei* gives a report concerning *Helias Artista.* and refers to the enlightenment of John (Chaps. 6 and 9) and considers the time has come for the partial destruction of mankind by war. Also by release of the angels at the borders of the Euphrates and subsequent spreading of the plague, when fully one third of mankind has perished and the victory of the

Lamb has been achieved. Then is the order restored, the face of the Church will be lifted; then the world will be under Christ's rule and the Jews converted.

Chapter III

The Preparation of the Wine Spirits of the Adepts

*(Spiritus Vini. Philosophicis.
Spiritus Vini Lulliani)*

The original description of this is from Raimund Lull in his *Libr. de Quinta Essentia* and Weidenfeld starts with that. As:

You distill the best red or white wine — *Vinum rubeum vel alhurr* in the ordinary way to obtain *Aqua Ardens*. This will be rectified three times and kept so that the burning spirit does not evaporate. The unmistakable sign is that sugar which has been soaked with it, when brought into a flame, burns just like brandy. When the water is prepared in this manner, you have the material out of which the quintessence will be drawn. You put it into a circulating recipient seal it hermetically, and place it in horse manure where the heat remains as a constant. It is important that the heat does not decrease, otherwise the circulation (digestion) of the water is distributed and not maintained, which is desireable; if however, a constant heat is used, the Quintessence will separate later in the digestion process, which is visible by the line that separates the upper portion, i.e., the Quintessence, from the lower portion. After a sufficiently long digestion, the recipient may be opened, and if a wonderfully pleasant aroma emerges, one which cannot be compared with any other pleasant smell in the world, and one which simply assails everyone, THEN you have the Quintessence. If this does not occur, the recipient must be put back and left until this goal, as described, is achieved.

This *Aqua Ardens, Spiritus Vini Philosophici* closely resembles the ordinary wine spirit and this is why it has not been recognized. It differs, however, inasmuch as in the process of continued distillation, and oil will separate and swim on the surface, which will not happen for the other material. It is the basis, the beginning and the end of all dissolvents of the Adepts. In its simplicity, it is perhaps the weakest but when combined with other materials, it is the strongest *menstruum*. It appears in two forms, one, like ordinary wine spirit and mixable with water, the other, as an oil on the surface. It is always the same thing, the difference being only in the purity and fineness.

Lull's method is actually correct but it comprises only a part of the process which is explained in other "recipes" such as I have compiled from Weidenfeld. We would like to use this opportunity to explain the word *"MENSTRUUM"* according to the Weidenfeld definition. For a long time, this word held civil rights in chemistry! The adepts have always used the allegory of Creation to veil the preparation of the Stone of Wisdom. Just like the embryo in the uterus is nurtured and gradually formed to maturity by the retained menstrual blood, the secret dissolvent constitutes, like the menstrual blood, the means to nurture and form the chemical child, the Philosophical Stone; therefore, they called it Menstruum, the name which has subsequently been passed onto all solvents.

COCIUM VINOSUM PARISINI

(page 128)

After the distillation of the *Aqua Ardens* and the phlegm, a black substance like melted pitch remains. This is washed out with the phlegm, mixed with the alcohol, digested and distilled, which is repeated with fresh alcohol until the residue is quite dry. The distillate is called *Spiritus Animatus*. This in turn is poured onto the residue, in increasing amounts, and digested until it is totally absorbed and the residue is of a white color. Following proceeds the sublimation. The sublimate is found to be clear and white as a diamond is. It is placed in a water bath where it turns to a liquid; then, the excess water is distilled off. Now, it is distilled four times with the first alcohol, using always fresh portions of alcohol. The distillate then is digested for 60 days. The success of the work can be determined when, on the bottom, a residue has formed that is similar to that of fresh, healthy urine. The Quintessence is then separated and is found to be so clear and light, that its presence in the glass might be doubted! Keep it in a cold place, well-sealed.

This is explained in a slightly different way in Weidenfeld on page 134 follows.

COELUM VINOSUM LULLII

Here the *Aqua ardens* is poured directly onto
the black residue; digested, the *Aqua animata*
developed and the oil is distilled off at higher
temperatures. The residue is calcinated until it
turns white. Then it is soaked with the *Aqua
anirriata* four times and sublimated. The shiny
sublimate is mixed with the *Aqua animata* and
distilled once, whereby the salt is transformed too.
The distillate is placed in digestion for 60 days
and turns into a pleasant smelling quintessence,
clear and light like a star. On the bottom you find
a salt, like in the urine of a healthy young man.

Another explanation is found on page 138.

SAL HARMONIACUM VEGETABILE PARISINI

The black residue is washed out with phlegm until it is white and shiny like a diamond. Then it is distilled with *Aqua ardens* in mild heat until the veins disappear; then the receiver is changed and the phlegm is extracted with higher temperatures. Like before, the residue is again distilled with the *Spiritus ardens* until it turns white and does not smoke on a glowing plate. Then it is repeatedly saturated with the *Spiritus animatus,* digested and all humidity is extracted. When a piece of it is placed on a glowing plate and mostly evaporates in smoke then sublimation follows. This is the *Sal harmoniacum Philosophorum.*

SAL HARMONIACUM VEGETABILE LULLII

The remaining thick substance, like poured pitch, is treated with *Spiritus ardens;* thereupon, first the *Spiritus animatus,* then the phlegm and finally the oil are distilled until they are dry and won't fume on a glowing plate. Then the eighth part of *Spiritus ardens animatus* is distilled as many times until it becomes volatile, which you can see when it completely goes up in fumes when placed on a glowing plate. Now it is twice sublimated, then dissolved in *Spiritus ardens,* distilled, and the distillate is digested in 40 - 50 days into a pleasant smelling liquid.

SAL HARMONIACUM VEGETABILE LULLII

TERRA FOLIATA

The spirit is distilled from the *Succo Lunaria (Vino philosophico)* with the mild temperatures of a single lamp until veins appear. This indicates that it is distilled. Now another recipient is attached and the second water which still contains some spirit is distilled until pure, tasteless water passes over. The black residue is then calcined. This may not be done with fire, as the Sophists say, but only through its own spirit. Therefore, the second distillate *(Aqua ardens* mixed with phlegm) is poured on it, dissolving it immediately. Then it is distilled over a lamp until the veins appear; that is when another recipient is attached and the distillation continues. This is repeated until it is like a black powder or until no more phlegm passes over, and the last water's smell and taste are as strong as those of the first water. The residue is now treated with the fourth part *Spiritus ardens* at low heat, until it is white as snow; then it is put on top of the fire where after 30 hours a magnificently white powder as light as silver, settles along the walls. This is *Terra nostra foliata.*

SAL HARMONIACUM LULLII

The black residue is extracted with the phlegm and
this process is repeated many times until it keeps
its color; after the evaporation an *Oleum vegetabile*
remains. The dry residue is distilled three times,
with *Spiritus ardens*. On the black calcined residue
you pour the *Oleum vegetabile;* let it digest for 10
days in the ash—bath; then you add the *Spiritus
animatus;* distill it away, and subsequently the *Sal
volatile* is sublimated.

COELUM VEGETABILE CIRCULATUM LULLII

You digest the *Spiritus ardens* in a flask with its neck turned downward until it floats lightly and clearly like oil on top. Then you open the seal with a needle, let the impurities flow out, and quickly turn it around. This is the *Spiritus ardens circulatus* with a most pleasant smell. The black residue is extracted with the phlegm; it is calcined and soaked with the *Spiritus ardens circulatus*. If a portion of it almost completely evaporates on a glowing plate, then the *Sal volatile* is sublimated, then dissolved in *Spiritus ardens circulatus,* and digested, and thus the quintessence is maintained.

MERCURIUS VEGETABILIS LULLII

The pitch-like residue is extracted with phlegm and distilled leaving the *Oleum vegetabile*. Onto the black residue your pour *Spiritus ardens* and distill it; then it is calcined in the reverberatory furnace, and the salt is extracted with the phlegm. Onto that *Spiritus ardens* is poured and distilled until it passes over unchanged. The thusly condensed salt is digested with the *Oleum vegetabile* and then distilled.

AQUA VITAE RECTIFICATA LULLII

The first *Spiritus ardens* still contains some water and a linen soaked in it ignites in a flame; however, does not burn: after repeated rectification, the soaked linen will burn up completely. On the pitch—like residue you pour *Spiritus ardens rectificatas,* distill, and then the *Oleum vegetahile* results. The black residue is distilled with the last *Spiritus ardens;* then it is calcined in *"Rerecherio"* and distilled seven times with the latest won alcohol; it is then called *Aqua Vitae rectificata.*

The complete process is as follows:

The *Vinuin rubeum vel album,* the secret philosophical wine, is distilled in the usual manner. The spirit thus obtained still contains water, and a linen soaked in it will ignite but not burn. With repeated rectification, it becomes so strong that a linen soaked in it will completely burn.

The *Spiritus* passes over in veins, and when those disappear, the collector is changed and the phlegm is distilled out; after the first distillation it still contains some spirit and it is kept for future use.

The spirit is put in the heat of horse manure to digest until an extremely pleasant smelling oil separates on the surface, which constitutes the quintessence. Lull obtained it with light blue color; others with a yellow color.

After the spirit and the phlegm have passed over during distillation, a black substance like melted pitch remains. This is extracted with the phlegm of the first distillation, until it does not change color anymore. The discolored portions are combined and distilled off, leaving an oil.

The residue extracted this way is calcined. This is done in different ways. In the method on p. 143, Lull says the calcination may not be caused by strong heat, but only by the *Spiritus ardens;* on p. 170 and 172, however, he says that it is done in the reverberatory furnace.

In the methods on pages 138 and 168, it was white through the distillation with the phlegm, but on page 143, it is still a black powder after the same treatment, and on pages 161 and 172, it remains black after being treated with *Spiritus ardens.*

The thusly prepared residue is digested and distilled with *Spiritus ardens* in varying conditions as many times until it is fully saturated and white, and the spirit passes over unchanged. The sign is that a portion placed on the red hot plate will not fume anymore. Then it is distilled repeatedly with *Spiritus ardens* until it becomes so volatile that when placed on a red hot plate it evaporates completely or to a large extent.

When it is prepared thus far it is sublimated. The sublimate is clear and light like a diamond. It may be used for the preparation of the *Spiritus Vini philosophici* by repeated distillation with the *Spiritus ardens,* whereby the *Sal volatile* passes over. The distillate is kept in digestion for 60 days during which time it turns into the pleasant smelling quintessence which is so clear and light that it can hardly be seen; the sign is a residue that deposits at the bottom like the urine of a healthy young man.

SAL TARTARI VOLATILE

Von Helmont established the reputation of the high medical power of the volatile alkaline salt; in his description he says (page 377 of the German edition): if impurities are found in the first processes you must add dissolvents; if they persist, however, then you need the volatile alkaline salts which cleanse everything like a soap. It is certainly astonishing how much a tartar salt, when volatilized, can do because it cleanses all veins of impurities.

(On page 1142). When the fire—resistant salts are volatilized, their power becomes similar to that of the great medications. They proceed up to the entry of the fourth digestion process and dissolve all blockage.

(On page 351). The first one is the alcahest. If that cannot be obtained, then you must learn at last how to volatilize the tartar salt so that you can prepare your solutions with their help.

(On page 329). The tartar salt (weinstein salz) can be completely volatile; it rises at times liquid and often like a sublimate. This salt has been

proven in tests even though this measure is less
known.

De le Bo Sylvius, in his time the pride of the
University of Leyden, and the founder of a new
chemical-medical school, also knew the *Sal Tartari
volatile.* The school, however, with its doctrinary
exploitation of the consequences of the system, destroyed this reputation again, which should serve as
a warning to us not to become the target of the
opponents working in the form of *Doctor opiatus.* The
solid tartar salt (Laugen salz), he says on page
850, may be volatilized by cohabitation with a volatile spirit. Such a volatile tartar salt rises and
sublimates at medium temperatures. Such a volatile
tartar salt (Laugen salz) is only granted to the
artists with diligence and patience; not to others
who avoid a long working time. Such a salt has great
powers.

Helmont's high regard consisted of an inducing
invitation to experiments, which, however, did not
give worthwhile results since they were done with
ordinary wine spirit and not with the wine spirit of
the experts.

The inventor of that substance is Raymond Lull,
and Weidenfeld gives us the method.

SAL TARTARI VOLATILE LULLII

Tartar salt (weinstein) is calcined for 3 days until it turns white; then it is dissolved in the not yet rectified *Spiritus Vini philosophici,* heated for 2 hours in the ash—bath, and the solution is drained. The residue is again calcined, repeatedly treated in the manner until it is totally dissolved. The solutions are distilled in the water bath and the distillate is reserved. The residue is placed in the ash-bath for 3 hours to remove the phlegm. Then the reserved water is poured onto the residue and distilled. This is repeated until the whole substance turns into an oil.

Further treatment now follows. On this oil you pour 6 times as much *Aqua Vitae rectificata,* digest it for several days in "*balneo,*" and distill it at low temperatures in the ash-bath until no more veins appear. As soon as the veins disappear, you take off the collector with the distillate and close it tightly: for now develops the *Spiritus animatus,* which is extracted at higher temperatures. The residue is ground, digested with four parts *Aqua Vitae,* and then distilled. Of the residue a small portion is placed on a red hot plate, and if it glows like wax without smoke, it is a sign of success; if that does not occur, the process has to be repeated until that sign happens.

On this residue your pour ¼ *Spiritus animatus* and let it congeal in the "Balneum," after which you evaporate the phlegm, which acts like pure water. Then you add fresh spirit and repeat that until the residue has absorbed all the alcohol, a sign of which is that if you place some of it on a red hot plate, most of it will dissipate in fumes. Now the substance is ready for sublimation, which is done at higher temperatures. The sublimate serves to fortify the *Spiritus Vini philosophici*.

We know that the potassium carbonate as such cannot be volatile, which means that the *Sal Tartan volatile* is no longer a potassium carbonate, but a potash salt treated with *Spiritus Vini philosophici*, and thus transformed and whose composition remains to be explored.

Chapter IV

Explanation of the Secret of the
Wine Spirit of the Adepts

In the second part of the mineral solvents Weidenfeld sheds light on the secret of the *Spiritus Vini philosophici,* which explains it to an extent. Different descriptions in that regard combined, provide the following information.

The secret material for the philosopher's stone which has been hidden behind many names (*prima material Lapidis)* is calcined and dissolved in distilled wine vinegar. The solution is evaporated until it takes a thickness of a gum. From that, first you distill a tasteless water with gentle temperature; when white vapors appear another recipient is attached and the *Aqua ardens* is obtained. This water has an extremely strong taste and at times a stinking smell, therefore it is called *Aqua foetens, Menstruum foetens*. If the distillation continues at higher temperatures, a red vapor and finally red drops appear. You let the temperature gradually die down and keep the distillate in a tightly closed glass so that the volatile spirit may not disappear.

The residue in the retort is black as soot; it is strewn on a stone and ignited at one end with glowing coal. Within half an hour, the fire spreads over the whole residue and gives it a yellow color; then it is dissolved in distilled vinegar, evaporated to a gum—like consistency, and then distilled. This is repeated often until the biggest portion is reduced to liquor. This liquor is poured into the first distillate where it digests for 14 days and then is distilled. First appears the *Aqua ardens* on top of which floats a white oil. This distillate is rectified seven times vntil a linen soaked with it and ignited will burn. A yellow oil remains which is distilled at stronger temperatures.

The sublimate in the neck of the retort is allowed to flow onto a steel plate in a cold place; to the filtered liquor you pour some *Aqua ardens,* whereby a green oil separates on the surface, which is taken off. Now the distillation continues; first comes water, then a thick black oil. As soon as white fumes appear, another collector is attached and the whitish distillate is extracted with medium temperature until a thick oily substance, like melted pitch, remains.

This black substance is treated further until
the residue is exhausted; but more explicit
explanation is unnecessary.

Ripley says that the *Menstruum foetens* derived
from the aforementioned gum contains 3 substances:

1)   the *Aqua ardens* which burns like ordinary wine
spirit when ignited;

2)   a thickish white water. The *Lac virginum* of the
adepts;

3)   a red oil. The blood of the green lion of the
adepts.

He says that nobody ever spoke this openly
about it and he fears the wrath of God and the
experts. With that, says Weidenfeld, he revealed a
big secret of the trade. The experts in their practical directions did openly discuss and teach the
use of the *Vinum philosophicum,* but how it could be
prepared was kept secret. Ripley is the first and
only one who says that the key to all of chemistry
lies hidden in the *Menstruum foetens* with its *Lao
virginum* and the *Sanguis Leowis.* When kept in mild
digestion for 14 days there results the *Vinum rubeum
et album Lullii,* and to confirm this he adds that

74

from the *Menstruum foetens the Aqua Vitae rectificata Lullii* are prepared.

The source material, the *prima materia,* has different names to hide the secret. The experts worked some in metals, some in metallic salts and ores. The *Leo viridis* name comes from its green solution; it is dissolved in sulfuric acid for cleaning, and it yields tungsten yellow crystals during evaporation. The prepared prime material is then calcined until red, thus eliminating the acid; then it is dissolved with distilled vinegar and thickened to a gum-like consistency, the distillation of which provides the *Spiritus Vini philosophici.*

The facts that:

1)   the prime material, calcined until red, is dissolved in vinegar forming an acetate salt;

2)   the black residue in the retort can be ignited and smolders, a characteristic of acetate salts;

3)   the distillation provides a spirit that burns like ordinary alcohol and it also provides a volatile oil, indicate clearly that nothing else is being taught than the preparation of the acetone.

For better understanding it might be good to give Wiedenfeld's presentation of the nature of the *Spiritus Vini philosophici* according to his remarks given here and there.

The *Spiritus Vini philosophici, Spiritus Vini Lulliani* is the basis, the beginning and the end of all solvents in the secret chemistry. It is, depending on the various degrees of its power, the weakest one or the strongest. It is the weakest when it dissolves by its mere oiliness (unctuositas) only the fatty parts (partes unctuosas) of the vegetabilia, while leaving everything else undisturbed: it becomes the strongest one, the more its oiliness is moderated by the acids, thus homogenizing it with dry fatty materials and the pure acids. Due to this homogenity, the solvents of the adepts differ from the ordinary solvents in as far as they stay with the dissolved materials and together with them are transformed into a third (therefore a chemical solution).

The *Spiritus Vini philosophici* appears in two forms, either as an oil floating on the top, or as ordinary wine spirit that mixes with the phlegm, but that may be separated by simple distillation, and that when ignited after rectification, will burn: they are, however, not two, but only one, different only in fineness and purity. With the ordinary wine spirit it has in common that during distillation the phlegm goes first, which is separated in the same manner.

The *Aqua ardens* (the first distillate) loses its watery form and concentration during distillation, and finally segregates an oil floating on the surface. This oil is dried through continued distillation and sublimated like a volatile salt through strong temperatures.

The oily *Spiritus Vini philosophici* extracts only the oily essences of the vegetabilia, and divides through simple distillation into 2 different parts, 2 oils or fats, of which one is the essence, and the other is the body; by further digestion with *Spinitus Vini philosophici* they are irrevocably reunited, whereby the spirit not only increases, but it is also modified to better dissolve dry material by the dry *(arida)* components of the oily body.

The preparation of the *Spiritus Vini philosophici* is the most secret, most difficult and most dangerous work in all of the secret chemistry.

The *Menstrua vegetabilia* prepared with it are sweet, without any corrosives, and dissolve the materials mildly.

There are different ways to prepare the oleum or the *Essentia Vini* from the *Vino philosophico*. Depending on the methods used, there are differences in preparation time as well as in smell and color.

Only when a mineral or metallic material has been dissolved in it, is the smell that pleasant.

This first of all dissolvents serves also as a medication with the name *Essentia or Specificum ad vitam longam.*

According to the rule of the *Chemia adepta: Essentia essentiam conficit,* become therefore easily essences for other material prepared for medical use and are given then special names. Paracelsus for instance names these: *Alcohol Vini de Pino, de Chelidonia, Essentia Melissae etc.*[3]

The *Spiritus Vini Philosophici* without condensing has no dissolving power over the dry materials *(arida).* This condensing is the secret of

---

[3] Paracelsus' descriptions are only vague and incomplete, as was his way, but Weidenfeld makes them somewhat more understandable. — HWN

the trade, difficult and tedious. It is.best done with honey, sugar, manna, salts and herbs and volatile salts. The highest degree of condensing and effectiveness is achieved by combining it with acids and mineral salts, whereby the *Menstrua mineralia* are formed.

Take the *Essentia Melissae de Vita longa C. III. C.5.* The *Melissa* is digested for 40 days; then through cohobation, the two components are separated, creating the *quinta essentia,* which is the elixir of life. After extracting the alcohol and its separation, then the *Vinuni salutis* appears with which the philosophers have been working for centuries without any results. Many of those he says mockingly, who have followed Raymund, have used quite some barrels of wine in order to find the *quinta Essentia Vini,* but they got nothing but a *Vinum adustum,* which was used improperly instead of the *Spinitus Vini.* The fact that Paracelsus, however, did know the *Spinitus Vini Lulliani* and that he also used it can be taken from the same des- cription of the *Spiritus Vini (de Vita longa, C. III. C. 9.).* The wine is digested for 2 months in horse manure; then you see a very thin and pure layer like a fat on the surface, which is the *Spinitus Vini;* everything underneath is phlegm. This

fat when digested alone and separately is highly
effective for longevity.

   The *Spinitus Vini philosophici* is dissolved by
the acid with the strongest heat, and therefore it
must be made certain that not too much is poured at
a time, and that the distillation has to be done
with extreme care. The *Menstrua* are stronger
depending on how often they have been extracted by
the acid which weakens through dissolution; they are
called *noctra* or *philosophica,* or *Acetum
philosophicum, Aqua fortis nostra, Spiritus
Vitrioli, Salis noster, etc.*

   The *Menstrua mineralia* have a stinking smell, a
corrosive taste, are mostly milky and turbid, and
dissolve materials with extreme power and heat;
since they have the *Spinitus Vini philosophici* as a
base, however, they are as permanent as the latter,
but not immediately the first time, but after
repeated cohobation. Continued cohobation will make
them sweet, and when the acid is taken away again,
it turns back into what it was before, i.e.,
Spiritus Vini philosophici. The acid cannot destroy
the nature of it, but only helps reduce the size of
the particles through permeation, thus making them
easier to dissolve. The *Menstrua that* are not
prepared with the immediate prime material of the

Spiritus Vini philosophici, but with the alcohol and acids which have been cleansed by circulation and distillation, stink less and are less milky, and the *Acetwn philosophicum* prepared in this manner is very light.

The *Menstrua mineralia* do not only dissolve the metals, but also make them volatile. The experts used them to speed up the work, and Paracelsus rightfully took over the monarchy of the arkanes by not only adding a final touch to these shortcuts, but also by introducing these *Menstrua mineralia* with such talent into medical application, that his students could hardly hope to improve it any further.

Chapter V

The Acetone

The wine spirit is chemically always the same, but technically and physiologically it is different depending on its preparation from grain, rice, potatoes, wine, etc.; the same holds true for the acetone depending on the various bases of the acetate salts; that is why I will give the individual descriptions as follows:

1. Acetone from Zinc

(RESPUR FROM *MINERALGEISTE* – p.116)

Zinc flowers were dissolved in a distilled wine-vinegar, then filtered and evaporated to oil consistency; when removed from the fire, the substance coagulates forming a salt. This was put into a glass retort and distilled. First it was flowing, then it started to pass over like a secret wine spirit in fine veins, however tasteless; then followed a thick and reddish water. With strong heat the whole substance swelled up and from it rose a ghost-like (spirit) snow which deposited in large amount, a thumb's thickness, and which fell down in some parts due to its volume. That which penetrated the receiver's paper seal had a smell as pleasant as Bernhard von Trevis has described it in his "left-out word,[4]" and I was quite surprised. After everything had cooled off, a thin coat with silver-white shine and prettier than Oriental pearls appeared all around; it could be touched with the fingers and had a smell like camphor.

Glauber *(Furn. Phil. 2 Th. p.* 99) also mixes the zinc acetate with sand, distills, notices however only that first a tasteless phlegm, then a subtle alcohol, and finally a yellow and red oil pass over.

---

[4] "Left-out word," Verbum dismissum, is the name adopted by the adept of the secret material which is not named in it, and it is therefore noteworthy that Respur often names the Zinc, thus explaining the secret Fontina Bernbardi, his solvent. – HWN

2. ACETONE FROM LEAD ACETATE

The experts worked much with lead, and Basilius Valentinus says that the Philosopher's Stone has its origin solely in lead; he also says that from the lead sugar a red oil is prepared, but he gives no further direction[5].

The first clear description can be found in Quercetanus (*Pharmacopoea p. 553*). The important thing in this description of the wine spirit of the experts is that for the first time lead is definitely mentioned, while the experts had always kept us in the dark about the basis. The lead sugar gives a highly burning water during distillation, which has a stronger taste than wine spirit. The recipient is filled with white fumes and finally an oil as red as blood follows.

From this *Liquor ardens,* which ignites faster than wine spirit, a spirit which is even more etherous may be separated with low fire. The black residue is calcined, the salt extracted and crystallized. Then it is soaked with the etheric spirit that much, that a fume rises when you throw it on a glowing plate. Through sublimation you

---

[5] lead sugar = lead acetate - HWN

obtain the *Terra foliata philosophorum,* which has a shine stronger than that of Oriental pearls.

When the red oil is added to this *Terra foliata* and combined with it through repeated cohobation and distillation, there results the true solvent of nature and the quintessence of magnificent power; this quintessence is the true and living, clearest source in which the vulcan washes Phobus (the gold), and cleans it of all impurities and creates the means to fortify the strength of life, improves everything weak, and renews the power of youth.

OLEUM SATURNI LULLII
(From "Fire and Salt" - Blaise Vignere, page 146)

Silver litharge is boiled with distilled vinegar and the solution is evaporated. The salt obtained is filled into half of a retort and the excess moisture is extracted using a gentle fire. As soon as white vapors are observed, a large recipient should be attached and the fire gradually increased, which will cause a small flow, like a milk—white oil, to rise in veins, which dissolves in the recipient like a hyacinth—colored oil, and whose smell resembles the spike oil. This is the secret oil of which Raimund Lullius did not say much more than: "*Ex plumbo nigro extraditur Oleum*

Philosophorum aursi colons vel quasi, et sicas, quod in mundo nihil secretius eo est".

On top of the residue in the retort, you can place glowing coals and it will catch fire like dry gras. It can be dissolved again with vinegar (the ash) and the above process may be repeated.

You take this oil, which Raimund Lullius calls *his wine,* and put it in a small flask over a water—bath, so that the spirit rises in small threads like the wine spirit. You distill until large drops appear in the helm, which is an indication that the rest is only phlegm. This is removed and at the bottom remains a precious oil which dissolves the gold and is good for all internal and external wounds; it is even a potable gold. Therefore, Ripley (p. 89 of the preface to his *Twelve Gates)* says: A gold colored oil is extracted from our subtle red lead, of which Raymund says that is is more precious than gold, because when he was near death in his old days, he prepared from this the *Aurum Potabile* and he regained his strength.

The burning water which also passes over is far more combustible than gun powder, and it dissolves silver into fine crystals which can be melted with a

lamp fire, and which like the silver stands up to
all tests.

AQUA PARADISI JOHANUS HOLLANDI
(Opus Saturni C. 12)

Lead sugar, completely purified, is distilled
first with gentle and later with stronger fire until
the material passes over red as blood and thick as
oil and sweet like sugar with a heavenly smell. The
residue is treated with distilled vinegar and in the
same manner distilled, and this is repeated until
everything is distilled into a red oil.

SPIRITUS ARDENS SATURNI
(Beguini Tyrocyn chem. 1616. C . 4. p. 139)

You keep the lead sugar for 1 month in gentle
heat so that it is in constant flux and then it is
distilled from a well-luted retort. The smell is so
pleasant that it fills the whole room and exceeds
the pleasant smell of all vegetabilia. On the distillate floats a yellow oil, and a blood-red oil
settles to the bottom. Through repeated distillation
the phlegm is separated and the pleasant smelling
spirit is saved.

SPIRITUS SATURNI
(Agrikol. Anmerkung zu Popp's chem. Ars. T. i p. 422)

Lead sugar is digested with good *Spiritus vini* for 4 weeks in the steam bath; then the spirit is extracted and a nice, thick liquor remains. This liquor is mixed with pure sand and *pen gradus* distilled from a retort, giving us a nice white spirit and a nice yellow and a red oil. The alcohol and the oil must be rectified together from a glass retort in a steam bath. First the spirit passes drop by drop; you see no veins or stripes; then follows a yellow oil; another recipient is connected and well luted; otherwise the fine vaporous aroma, more pleasant than amber and musk, will be lost. If the yellow oil is over-distilled, the phlegm will appear in many snow—white streams; then another recipient must be connected and all phlegm passed through. There finally comes a nice red oil, whereby a higher temperature is necessary because it is heavy and does not rise easily.

QUINTA ESSENTIA SATURNI
(Agrikola 1. p. 242)

The process is the same as above. The spirit and the oil are individually rectified one more time.

The black residue in the retort is calcined with high temperatures, until it is snow—white; then it is dissolved and crystallized with distilled vinegar. This salt is kept to digest with the previously rectified spirit for 8 days in a steam bath. Then it is distilled, whereby most of the salt will rise. The distillate is poured back onto the residue; then it is again digested and distilled, and this is repeated until the whole *Sal volatile* has passed over (in gestalt) in the form of spirit. Now the rectified red oil is added, whereby the two are inseparably mixed and make an extremely tasty medicine.

RED OIL FROM LEAD
(Experimentirte Kunststucke. 1789. Th. 1. p. 150)

Lead sugar, from a glass retort filled up to one quarter, is distilled in the sand cupel. At first you get a very sour spirit; after that the recipient is changed and the temperature raised. Then follow brown, stinking drops which must pass until all humidity disappears. During this time the substance in the retort will have risen somewhat and will appear black and layered like an empty wasp nest. The temperatures are increased and ruby red; pleasant—smelling sweet drops appear. During the first experiment the retort had ruptured so that very little of these drops could be saved, but the beautiful balsam odor filled the house and the whole street.

SPIRITUS ACETL. ARDENS
(Charas Phamacop. royale p. 775)

You distill lead sugar at first with gentle and later with stronger heat. The distillate is rectified with mild temperatures so that first the burning alcohol passes over, followed by the phlegm, leaving a purple red liquid which you very inappropriately call *Oleum Saturni,* and which does not have a very strong acidity.

The distillation of the lead acetate was disappearing from chemistry until in more recent times Chenevix picked it up again, giving cause for further studies of the acetone with his "*Esprit pyroacetique.*" Mainly the acetone itself was studied and little attention was paid to the other products than had been the case in the old chemistry, when extra care, patience, and persistence were used which is why Weidenfeld calls the preparation of the *Spiritus vini Lulliani* the "most difficult task."

3. ACETONE FROM COPPER

(Spinitus Aeruginis Basil. Valentui. p. 834)

Pure, crystallized verdigris is calcined until it starts to become reddish. Then you take 2 parts of it, 1 part pebble stones, which have been cleansed repeatedly in vinegar, rub them together, fill them into a fogged up glass retort, attach a large and well luted collector, apply medium fire for a while day and night, and then increase the fire for a day and night, so that at first greenish—white alcohol, and after a long time, occasional red drops will appear. The fire has to be maintained until everything has passed over. The distillate is mildly rectified in the water bath so that the phlegm disappears and a heavy red oil remains at the bottom.

SPIRITUS AERUGINIS

(Zwelfer Appendix ad Animadvers. as Pharmacop. 1685. p. 51)

Spinitus vini rectificatus is 2 or 3 times distilled over crystallized verdigris; then the crystals from a fogged up retort are distilled in an open but gentle fire until all spirit is passed over and subsequently rectified.

Zwelfer, moved by his conscience, gave away the secret of this spirit and he also praised its chemical and medical powers. He compared it with the *Liquor Alcahest* because when these substances were gently dissolved, they could both be retracted with also identical strength; he recommended them especially for the dissolution of pearls, corals, and crab eyes, as well as for the preparation of the *Tinctura ex Vitro Antimonii* and *Tinctura Martis adstriingens*. This caused a bitter dispute spiced with Latin crudities, with Otto Tachenius, who said that the *Spinitus Aeruginis* is nothing but a distilled vinegar, and that Basilius Valentinus had already described it. Boerhave too declared it an acetic acid, however, the strongest that could be obtained from vinegar.

Chenevix's examination decided the matter; the *Spiritus Aeruginis* is not a pure acetic acid because it contains O_{17} a flammable acetic spirit because of its volume, and with this he justified Zwelfer. The two Derosnes distilled the copper acetate into 4 parts. The first part was light colored and had a faint odor; the second part had a stronger smell and dark color; the third one was darker yet in color and had a stronger smell of flammable acetic spirit. The fourth one was slightly yellow and contained a

rather large amount of flammable acetic spirit
(Thenards Chemie von Fechner IV. 1. p. 151).

4. ACETONE FROM IRON

(Agrikola 1. p. 418)

The blackish—yellow distillation residue of "Eisenvitriols" (ferrous sulfate) is repeatedly extracted with the help of. distilled vinegar. The solutions are evaporated until a green liquor remains. This is mixed with calcinated pebbles and then distilled. The distillate is digested for some time; then the phlegm is carefully extracted and the residue is twice rectified from the sand bath, resulting in a beautiful, sweet oil. According to Chevenix, the distillate of iron acetate contains flammable acetic spirit if you consider the volume.

5. ACETONE FROM STIBNITE

(Tinctura et Oleum Antimonii Rogeril Baconis)
(Deutsches Theatrwn chemic. III. p. 207)

Finely pulverized stibnite ore is individually placed in aqua Regia. As soon as it is dissolved, it is extracted and the residue is cleansed. This residue is digested with distilled vinegar for 40 days in a water bath when it gets a color as red as blood. The clear liquid is poured off and fresh vinegar is added and left to digest for 40 days. This must be done four times. The residue is discarded.

The solutions are placed together into a flask; the vinegar is distilled off and again cohobated, or if it is too weak, <u>fresh</u> vinegar is added, and after dissolution is distilled off again. The residue is washed with sweet water until all sharpness is gone. The substance which turns bright red is dried in the sunlight or in gentle fire.

To this red powder you add well—rectified *Spiritus vini* and leave it completely in a water bath for 4 days to dissolve. The solution is placed in a flask with a helm in a water bath; a receiver is attached and the alcohol is distilled at low temperatures. The alcohol is again added, again distilled, and this procedure is repeated until the alcohol rises in several colors over the helm.

That is when high temperature is needed to make the pure alcohol rise to the helm, and then drip into the collector as a blood-red oil. This is the most secret method of the wise for the distillation of the highly praised oil of *Antimonii,* a noble, strong, pleasant-Smelling and powerful oil.

The distillate, the mixture of wine spirit and oil, is placed in a flask with a helm, and the alcohol is completely distilled off in the water bath which may be determined by some drops of oil passing over. The alcohol will keep well because it still contains great power from the oil dissolved in it.

In the flask you find the blood-red oil which glows at night like coal; it is used for alchemical improvement of metals.

The wine spirit, the *Tinctura Antimonii,* is a very powerful medication. When you suffer from Podegra and take 3 drops dissolved in wine on an empty stomach the pain will subside; the next day follows a tough, thick, and stinking sour sweat, especially in the joints, and on the 3rd day, even without medication, it is an easy purgation it is just as helpful with other serious injuries.

QUINTA ESSENTIA S. OLEUM ANTIMONII
BASIL VALENTINE

(Triumphal Chariot of Antimony trans. by Kerkring p. 147)

Over very finely pulverized *Vitrum Antimonii* you pour distilled vinegar and under frequent stirring to avoid an assimulation, it is digested with gentle heat until the vinegar is tincted bright yellow. This is repeated until the vinegar does not color anymore. The solutions are filtered and the vinegar is distilled off in the water bath until it is almost dry. This has to be done extremely carefully because a heat that is too high spoils the preparation. The reddish—yellow powder has to be dried in the sun in mild temperatures. The powder is repeatedly washed (edulcorated) so that all acid disappears. Then it is finely ground in a lightly warmed glass mortar; then highly rectified wine spirit is poured over it up to 3 fingers high; it is digested and a bright red tincture results. This tincture is digested for 1 month and subsequently with a special method (*according to the Microscop.*

Basil. Valeni. p. 109 by mixing it with *Terra sigillata)* it is distilled over. It gives a lovely sweet medication in the form of a beautiful red oil, which is the *Quinta Essentia Antimonii.*

6. ACETONE FROM POTASSIUM

(Agricola II. p. 15)

Saturated potassium acetate liquid is kneaded into balls with pottery clay. These are dried in the air and then distilled from a retort. A strong but very lovely smelling spirit passes over, white as milk, which settles everywhere on the sides of the recipient, like a volatile salt. You let it stand for 24 hours, and it dissolves into a nice, clear, yellow oil.

Pott (*Exercit. chym. de Terra foliat. Tart. p. 152)* mentions that when he rectifies 1 part potassium acetate with 6 parts vinegar 3 times, during the 4th time half the salt has passed over and volatized.

7. ACETONE FROM ACETATE NATRON

Upon my initiative in 1840 the pharmacist Klauer took upon himself its preparation and reported the following:

4 pounds acetate natron gave 20 ounces distillate. The distillation out of the sand bath was completed within 3 days. The distillate was distilled in the water bath; first the acetone with some water passes over, the acetone passing over at 55°. The further stronger distillation provides water, acetic acid and some oil *(Metaceton)*. The residue is a dark brown oil of thick consistency, which dissolves easily in the acetone.

In order to keep the acetone water—free it is rectified over calcium chloride. 6½ ounces of water containing acetone, gained from 4 pounds of acetate natron, provided 4½ ounces of water—free acetone with the following characteristics:

1) A colorless thin fluid with a fine, penetrating smell, similar to the etheric acid (Essigäther).

2) Mixable with wine spirit and ether in all proportions.

3) Specific weight O_{708}

4) Easily ignited, it burns with a very bright and little sooting flame without residue.

The acetone yielded the following deposits:

1) With nitrate mercury oxide — yellow and "copios,"

2) With nitrate mercurous oxide — black,

3) With copper sulfate — blue,

4) With copper acetate — blue,

5) With ferrous oxide sulfate — greenish, later turning yellow,

6) With sulfate "Maagenoxydul" — reddish,

7) With acetate "Maagenoxydul" — reddish,

8) With chlorine-gold — segregation of metallic gold,

9) With ferric chloride — a gelatine-like substance,

10) With mercury chloride — a gelatine-like substance.

The acetone is combined with the 2 oils, and has been prescribed by me as a medication under the name *Spiritus Aceti oleosus*.

8. ACETONE FROM CALCIUM ACETATE

(Poterii Opp. p. 612)

The corals are dissolved in distilled vinegar; the solution is vaporized and the dry salt is placed in a luted retort. The phlegm is removed first with a low temperature; then with a different recipient the spirit is distilled over along with a small amount of red oil, both very pleasant smelling and bright red.

Quercetanus received 6 ounces of spirit from one pound of the coral salt.

In an experiment made in 1841 where acetone was prepared from calcium acetate, a product was achieved which differs from the one made from acetate natron. It did not smell as spicy but like pyrolignite; the taste was less fine; the empyrheumatic oil tasted, burned and had a stronger smell; therefore, it was not used as a medication.

In regard to the chemical characteristics of the acetone I observed the following: In Nov. 1861 in the pharmacy, I found a few ounces of an old test of *Spiritus Aceti oleosus*. It was colored yellowish and had its full odor. A sample of this when combined with sulfuric acid, turns dark red immediately, while this change in color occurred much later when the acetone from a chemical plant was used.

I placed the glass, which is closed with a ground glass stopper, on the back—stove. After 14 days parts of it had evaporated and a ruby—red oil had segregated on the surface. The latter smelled like acetone; the taste is bitter and lasting. It discolored litmus paper cinnabar-red, while pure acetone showed only a weak acid reaction after several minutes.

I added half an ounce of pure acetone which dissolved the oil instantly.

I returned the glass, still protected with the "gyps," to the back-stove. After some time with the easing of the "gyps," and the partial evaporation of the liquid, the ruby-red oil forms again and has remained since then, even when removed from the heat. When some drops are mixed with water, it separates quickly and settles to the bottom, but the taste of the water is bitter like the oil, and it smells like acetone.

Chapter VI

Medical Application of Acetone

I will not mention the general acclamations for its use against numerous illnesses from the alchemical literature, but I will limit myself to Kerkring and the experiments of Agricola, who specialized in this matter.

QUINTA ESSENTIA OLEUM ANTIMONII
BASILLIA

(Kerkring: Triumphal Chariot of Antimony p. 153)

A 21-year old lady with hydropsy had swollen up terribly. She took this medication twice a day. After 20 days she had sweated so much that her body had shrunk half an ell. She lost quite an amount of urine in that time and the sweat was quite wonderful. The medication does not have the same effect as other *Diaphoretica,* which with the first dosage causes sweating, but it only opens up the skin on the first day, causes mild sweat on the second day, and on the third day the sweat increases; only on the fourth and following days does one practically swim in water, so that finally the sweat drips through the bed onto the floor. This is, says Kerkring, when a knowledgeable doctor is needed, because the club of Hercules does not help much when it is not in the hands of a Hercules.

ACETONE FROM IRON

(Agricola 1. p. 425)

1.   Lung ulcers.

You prepare a syrup of 2 drachm per 10 ounces of syrup; of that often times a large amount of hazelnut is placed on the tongue. It helps with the cough; increases the phlegm and makes the breathing easier.

A 36-year old man was suffering from a severe cold, heavy chest oppression with the danger of suffocation. He had tried many medications without improvement. When he used the chest-syrup much pus disappeared; he took it for one month and his health was completely restored.

A 6-year old girl had been coughing up blood and pus for 2 years and she was all consumed. She took the medication only 3 times a day and recuperated completely within two months.

2.   Against Poisonous Stings.

A young shepherd was stung in his left thigh while sleeping. The spot was brown and as big as a 3—pence, and it hurt very much. The following day the thigh was brown and swollen. Warm acetone was placed on it and after two hours the swelling and the pain were less; after a fresh application and two more hours the pain and swelling had completely disappeared and the boy could walk again.

3.   Panaritium

In the case of panaritium the application eliminates the pain within one hour and the sore opens up soon.

Agricola himself during a trip had an infectious sore, i.e. an isopod sore between two fingers, and he suffered a lot of pain. Several medications did not help. When he got home he applied the acetone; the pain disappeared so that he could sleep again and after a few days the sore broke open and healed quickly.

ACETONE FROM LEAD

SPIRITUS SATURNI

(Th. 1. p. 239)

1.   Against stinging in the spleen with distension, 6 drops in Extr. Filicis.

2.   Kidney injection, whereby the fat melts and emaciation follows. A farmer always had fatty urine, as if melted butter had been poured in it; at the same time, he felt a lot of heat in his back and his energy and body were diminishing; he was losing weight on his hops and was always feverish. He took 3 drops Spiritus Saturni in Aqua Plantaginis at night; after four times he was healed.

3.   Gonorrhoea virulenta. A noble man was suffering from this for some time; he felt immense heat and thought that nothing but Abscissio membri could help. Spiritus Saturni with Aqua Sambuci was applied and soon extracted the heat; at the same time injections with this were made; he was healed in 3 days.

When applied against panaratia, it helps very quickly.

ACETONE FROM POTASSIUM ACETATE

(lb. 11, p. 15)

1 part acetone

2 parts *Spiritus Vini*

½ part *Cl. Vitrioli*

are digested for 6 weeks until it becomes a lovely and palatable medicine for many sicknesses.

1.   It is a very strong invigoration for the stomach; 12 drops in the first spoon of soup. The stomach may be full of phlegm; this will divide the phlegm and lead it away without any other medications.

2.   It eliminates the stomach fever completely if a mild catharsis is needed, and especially if the ill person does not feel very cold or very hot. This fever usually lasts for some time because of the

persisting phlegm in the stomach; the phlegm also causes permanent headaches.

3.   Against the stone. A preacher was suffering from strong stone pain, and all medications increased the pain so that he did not want to take anymore. Agricola told him that this medication does not push the stone, but opens the ways and dissolves at the same time the stone in the kidneys, so that it will pass without pain. He took 10 drops every day with a spoon of soup, and after using it for one month the pain was gone and the urine very thick and turbid with bright red deposits.

4.   In the case of hot pestilent fevers, it is a powerful means and resists the poison because it forces strong sweating when ½—1 scruples *in aqua* or *aceto theriacali* is given. You also add 1—3 drops of *Essentia Croci* so that the heart is not overcome by the poison. It is especially suitable for children because of its pleasant smell and taste.

5.   Against early or not too long—contracted Podagra, every day 15 drops in *Aqua Ivas artheticcze*. It locates the problem area and causes pain there; that is when that area needs also external application.

6.   In the case of pain in hollow teeth, you take syrup 1 in warm vinegar into the mouth and the pain is quickly eliminated.

TINCTURA ANTIMONII THEDENII

(Remarks & Experiments for the Enrichment of Medicine & Medical Science)

(1782. Th. 11. p. 84)

Theden prepared his tincture according to the instructions of an alchemical writing as follows: 2 pounds stibnite are melted together to a liver with 6 pounds of potassium and saturated with 13 Berliner quarts of concentrated wine vinegar. The substance was evaporated until dry; then *Alcohol Vini* was added and distilled in the water bath. The wine spirit that had passed over was poured back onto the substance, again distilled, and this process was repeated 30 times, whereby the lost spirit was always replaced. Sixteen quarts alcohol were used up and hardly 2 pounds tincture were produced. This tincture was digested for 3 months in the ash bath, during the first month with one, during the second month with 2, and then with 3 lamp fires, leaving one pound of tincture.

He administered this medicine in the case of glandular blockage, externally as well as internally, and the effect exceeded his expectations. Eight—ten drops, taken daily, caused sweat, increased urination, and upon increasing the dosage, soft stool and mild laxation. It eliminated the podagral pains, helped with clogged intestines, but the most important thing was that it achieved in 3 cases the complete division of hidden cancers, and in 2 cases, it helped promising good hope for improvement.

In the 3rd part, page 269, we find Dr. Walter's observations from Lief land, according to which induration in both breasts caused originally by hardened milk, was completely healed.

The famous Wichmann held the medication in high esteem against "breast browning": the only patient whom he had the rare fortune of healing of this severe sickness had been saved through the use of medicine for ½ a year through two fontanels in the thighs.

Theden says nothing about the color, taste, and smell of the medication.

As undoubtful as the medical effects were, as doubtful were the views of the chemists in regard to these effects. It was called stibnite tincture; the chemical test showed, however, that it contained no stibnite; also the procedure was so expensive, complicated, and time-consuming that the production encountered many obstacles. Gren said: it is a solution of the leaves-earth in wine spirit; the few stibnite particles which it might contain are not worth the painful preparation; and according to Westrumb it was nothing but a solution of the potassium acetate which after the long torment, as he states sympathetically has turned combustible. Not without irony says Elfers that it can be found in those pharmacies where the pharmacists do not have much chemical knowledge, and in Tromsdorff's Journal it was said that this ineffective tincture deserves to be banned from the pharmacies--which later actually happened.

This shows how the chemists of those days were without knowledge of the breakdown of the acetate salts through dry distillation, which the alchemists of the 13th until the 17th century had activated with such patience and attention as the goal of their secret work.

It is obvious that in the long way that Theden went, a gradual breakdown of the potassium acetate with partial decomposition of the wine spirit in acetic acid is caused, and an effective medication is gained whose chemical examination will be the subject of today's analysis.

Chapter VII

My Own Observations
Concerning the Application of Acetone

Since 1840 I have used the acetone very frequently. Since it contained not only acetone, but also oils, if prepared according to the old way, I called it *Spiritus Aceti oleosus,* so it could be distinguished. The preparation was good, but it did not correspond completely to the description of the old chemists since it was lacking the praised pleasant smell, which might be due to the fact that the preparation used then let the substance mature through long and repeated digestion and distillation, like wine when placed in a place with moistened rowen is refined by the warmth within 3 months, as if it had been stored in bottles for 3 years. As the old rules show, it is a very delicate operation whose basic rule is "Eile mit Weile" (hurry with patience). The dehydration of the acetone through distillation over calcium chloride is chemically correct, but not the best for the medication. The pure acetone, like it can be obtained commercially, is not as strong, not in smell and taste, not in its medical effectiveness.

Generally, I have noticed:

1. The urine and stool get a terribly stinking
smell like cat urine and cat stool. I observed that
immediately at the start of my experiments with a
lady who had the flu and for whom I had prescribed:

R. *Spir. Acet. oleos. Drachm. 1.*
Aq. destill. Unc. II
Syr. Sachar. Unc. semis (ounces 1/2)
Ms. every 2 hours 1 tablespoon.

During the second night she urinated, creating
a stink that filled the whole room so that it had to
be aired out. In the hallway where the night urine
was placed, the stool smelled just as badly. The
stink continued as long as she was still on that
medication which, however, became soon dispensable
because of the improvement.

In the case of a tailor in the last stages of
consumption, the stink in the stool and the urine
appeared already after one-half Drachms after one
day. During the following days it became even worse
and appeared also in the phlegm.

In the case of a nervous, hysterical woman, 5 drops gave her urine the special smell within half an hour.

An old lady received:

R. *Spir. Acet. oleos. Syrup. dimid.*
    Aqua destill. Unc. duas
    Syr. Sachar. Unc. dimid.
    Ms. every three hours 1 tablespoon.

No change in the urine. After the medication was used up, I prescribed:

R. *Spir. Acet. oleos. Drachm. 1.*
    Aq. dest. Unc. 11
    Syr. Sachar.
    Muci 1. Gumm. arab. aa Unc. dimid.
    Ms. every three hours 1 tablespoon.

During the night the stool smelled terrible and this condition persisted for as long as she was taking the medicine.

All ill persons were surprised by this; they believed, however, or let themselves be easily persuaded that damaging, rotten materials were segregated from the blood, and I thought it best for the experiments not to give any further explanation.

The urine itself showed no changes; it was sometimes sour, sometimes neutral, sometimes light, sometimes turbid; in some cases there was an increase in urination.

2.   I noticed no effect on the sweat; where it appeared it was more a consequence of the development of the sickness.

3.   A visible effect on the nerves was noticed.

A policeman had had a severe case of *Meningitis spinalis* which had been treated with the hot iron and the strongest medicines. From this he had retained a "neuvalgie" of the neck with continuing convulsive shaking of the head, which was worse when he was in an upright position, whereby the head and neck were pulled toward the back between the shoulders, causing severe pain. Consequently, he could never sit freely, but had to have something to lean his head against. In August 1840, when at the age of 57, and after several years of sickness, he started treatment with me. Until the end of December I tried a homeopathic cure with high thinnings of *Belladonna, Nux vomica, Coeculus* without significant success. In January 1841 he started with *Spiritus Aceti oleosus*. Soon some improvement showed; after 3 weeks he could sit for 6 hours and play solo. At the end of February the improvement had progressed so that he could sit freely and walk around; however, the head was still trembling, but it was not being pulled toward the back anymore. At the end of March, he could take short walks when the weather was nice, and only longer walks posed a problem for the back; the trembling and shaking of the head had remained, giving a strange appearance. He had used the medication continuously for three months, and since he was content with his condition, and his situation did not allow any further medical expenses, his treatment was ended. He kept the trembling of the head until his death in 1860.

The hysterical woman, whose urine started smelling already ½ hour after 5 drops of the medication were administered, felt a pleasant warmth in her stomach, which rose toward her head, causing much relief; the nausea and ache near the heart improved. The positive effect remained for the following days; the cramping in the limbs decreased; the dizziness decreased; the sleeping improved. There was strong urination with a distinct smell.

A very nervous woman fainted for one hour. After 8 days when she still had not fully recovered, she took:

R.   *Infus rad. Valerian. Unc. 11.*
     Spir. Acet. oleos. Scrup. 1.
     Syr. Aurant. Unc. semis.
     Ms. every two hours one tablespoon.

In the evening she felt better and more alive, and soon recuperated completely.

4.   Rheumatism.

A child had rheumatic pain in the back of its head and neck with slight fever irritations. After three days the child was not better.

R.   *Spir. Acet. oleos. Scrup. 2.*
     Syr. Sachar.
     Mec. Gummi arab. aa Unc. semis.

After that, the child slept in the afternoon like after opium, and the pain disappeared during the following days.

A lady suffering from frequent pain in her face, felt the pain coming and took:

R.   *Spir. Acet. oleos. Drachm. dimid.*

Syr. Sachar. Drachm. tres.

Syr. Cinnamom. Drachrn. unam.

Ms. 3 times daily 1 tablespoon.

After that the pain disappeared, but her head was slightly numb because, as she said, the medication was too strong. The taste was all hidden by the juice, but to her it still tasted like creosote. Probably some of that is the thick oil, but since creosote is a pain—reliever, the medication must remain as is.

In some cases, it increases the pain first, accelerates the development of the disease process, and brings out the rheumatism like the homeopathic process.

A young girl had heart rheumatism. After two administerings of ¼ grain *Aurum metallicwn praecipitatum,* the heart was free on the second day; however, a toothache appeared which increased during the third day with pain in the head and ear of the left side. On the fourth day, no changes. *Spir. Acet. oleos.* 5 drops 4 times daily.

Fifth day less pain.

Sixth day - from early in the day, strong pain and aches. Ten drops per dosage. The whole night less pain. Steady decrease of pain, and on the ninth day, they were gone.

A woman had strong headaches which continued through the night. *Spir. Acet. oleos.* internally and externally, an application of:

R. *Spir. Acet. oleos.*
Ol. Olivar. aa Drachm. 1.
Tinct. Kahn. Scrup. 1. Ms. to rub in.

Subsequent relief and sleep through all the night.

Third day - Less pain, but in the a'fternoon a renewed, strong attack that lasted the whole night until the afternoon of the 4th day; then rest and a good night.

Fifth day - No pain, good night.

Sixth day - Only slight pain; that disappeared completely.

A young girl experienced chill and swelling of the gums. *Spir. Acet. oleos.* In the evening her gums were better, but the lip was swollen with stinging pain in the skin at the forehead and the temple.

Second day - Crysipelatose swelling of the nose, mouth, and cheek; all other pain disappeared. The medication without sweat and urine had obviously pushed the rheumatism onto the skin.

Fourth day — Everything is better. During the night after a good sleep, severe chills.

Fifth day - In the afternoon heavy sweating, beginning of menstruation, 8 days too early.

Sixth day - Good condition.

5.   In the case of feverish conditions, the *Spir. Acet. oleos.* causes too much heat.

A young female cook felt poorly for 8 days, and started having a headache in the forehead, stitches in the side, and fever. *Spiritus Acet. oleos.* 5 drops every two hours. During the night rheumatic pain in the face and teeth, while the head and side pains disappeared.

Second day — The previous pain returns, also in the night.

Third day - The whole morning shivering, in the afternoon heat and thirst. Constant hallucinating, everything appears larger and stranger; when she closes her eyes, a figure appears that looks like a man in a coat without a head, which frightens her; also pains again.

The medication was stopped.

After 8 days the fever was gone, but the rheumatic pain remained in the breast, and there was also a feeling of oppression.

R.   *Spir. Acet. oh. Drachm. dimid.*
         Aq. destihl. Unc. duas.
         Syr. Sach. Unc. unam.
         Ms. 1 tablespoon every 3 hours.

On the following days she had felt all well. The medication had caused strong urination.

A man had a podagrous infection of the left wrist and fever. *Spir. Acet. obeos.*

Second day — Pain in the knee and ankle of the left leg.

Third day - Additional pain in the right arm and elbow, urine with strong, bright red deposits, coated tongue, no appetite, in the evening higher fever with much thirst.

The medication had apparently an effect of overheating and was discontinued.

AN EXPERIMENT WITH PURE ACETONE

In February 1862, an old but vigorous lady of 75 years suffered from acute rheumatism in the shoulders and in he back, which was extremely painful. After the fever was lowered, on Feb. 26th, I prescribed:

R. *Aceton pur. Drachm. unam.*

Aq. dest. Unc. duas.

Syr. flor. Aur.

Muc. Gumm. arab. aa Unc. dimid.

Ms. one tablespoon every 3 hours.

This was pure acetone from a chemical plant since the previous *Spiritus Aceti oleosus* was no longer available in the pharmacy. The medication had the taste of acetone, but it did not taste unpleasant and it gave the stomach the feeling of warmth.

After the intake the patient, who had always been suffering from hard stool, experienced soft, mushy stool with a terribly bad smell, but the urine had no smell.

She took the medicine until March 2, i.e. for 6 full days. The mushy stool kept its awful smell. She was comforted with the explanation that the had smell was due to the excreted rheumatism substance, and that it was a good sign. She said, however, that the stink was unbearable, and since the rheumatism had not improved, the medication was discontinued. On the two following days the stool was still the same, and only on the third day was it solid and without the acetone stink.

The pain vacillated between less and more when other medication was used.

On March 16 I prescribed:

R.   *Infus. rad. Valer. Unc. duas. cum dimid. Aceton*
     pur. Drachm. unam.
     Syr. Aurant. Drachm. sex.
     Ms. 1 tablespoon 3 times a day.

The medication was given twice. After that she had felt somewhat stronger; the stool had its smell again, but the pain remaixied unchanged. Therefore on March 21st I prescribed:

R.   *Tinct. Spigeb.*
     Tinct. Rhododendr. aa Drachm. dimid.

Aq. Nuc. vomic. Unc. dimid.

Syr. Sach. Drachm. duas.

Ms. 4 times a day, 25 drops.

On the 22nd, she was completely pain-free, but she slept unusually long and deeply during the day and at night, and upon awakening on the 23rd, she felt like she was paralyzed in all limbs, and the mobility was only gradually restored.

For this threatening situation, she received *Ammon. carbon pyrooleos. gr.* 1 three times daily.

On the 26th, the paralysis improved and the pain came back. For resuscitation, I prescribed:

R.   *Infus. hb. Rorismarin. Unc. quatuor. Aceton pur. Drachm. irnain.*

Syr. Sach. Unc. dimid.

Ms. 1 tablespoon every two hours

and externally I rubbed on *Ung. nervin.* with *Lini'n volatile.*

On the 28th, there was increased improvement, light stool still without the specific smell.

On the 29th, the stool had once again the terrible smell. She was all upset that in such short time so much waster material had collected in her body because for days nothing similar had been felt. She must have become suspicious of the medication because she determined that she did not want to take anymore of it and wanted to let good weather improve the situation. I for my part was content with my observations and agreed. After she stopped taking the medicine she improved daily, but the rheumatic pains returned periodically, and only after a long time did she regain strength.

From this follows:

1. The *Aceton purum* gives the stool a stinking smell just like *Spiritus Aceti oleosus;* of course this is only noticed if the discharge is done in the room.

2. It does not change the urine.

3. It had no healing effect on the rheumatism like the *Spiritus Aceti oleosus;* therefore, it appears that the etheric oil is essential for the medical constitution.

ANTIPYRETON POTERII

Petrus Poterius, who calls Friedrich Hoffmann the *"medicorum sui aevi Principem"* and whose *"Opera practica el. chymica"* he considered so instructive that in 1698 he published them with his preface, used a fever medication which he called *Antipyreton,* and which he says is the only and most effective of all.

He describes 24 fever cases that had been treated with it. These are: *Febris ardens, F. maligna, F. bihiosa, F. hectica, F. tertiana simplex* and *duplex,* and *F. quartana simplex* and *duplex.* In some cases only one dosage, in other cases, 3 to 1 were given per day. The healing took in most cases surprisingly little time, mostly just after a few days; only in 2 cases did it take 10-14 days. It is remarkable that in the case of *Tertiana duplex* a case of splenic tumor existed, which in one case was not completely eliminated, and in the second case it even increased.

These observations caused me in 1844 to prepare this medication and to experiment with it in the following situations:

1. SUDOR INTERMITTEUS QUOTIDIANUS

An old lady who had suffered from rheumatism for many years and who had a podagra node at the wrist, caught a "catarrhalisch" (catarrhous) — gastric fever which ended after 3 weeks so that she could leave the bed. That is when every afternoon a period of sweating set in which at first lasted for 5 hours, and only gradually decreased. This sweating continued regularly for 7 weeks and the medications used brought no change; intermittently also rheumatic pains appeared.

This was the first case where I prescribed the medication. In the evening she took three drops of *Antipyreton*. After that, she had at first a feeling of comfortable warmth through her whole body, then a tickling in all limbs, then she fell asleep and woke up after 2 hours sweating, whereupon she fell asleep again and sweated. The sweat was stickier than usual.

Second day - She felt very good and strong. In the evening another dosage of *Antipyreton,* after itching in the face, especially around the nose, tempting to be scratched, then sleeping with general sweat.

Third day - In the evening *Antipyreton,* after that only mild itching in the face, but consistent itching in the breast with a slight cough and expectoration.

Fourth day - Fairly well, the podagra lump has shrunk noticeably. No medication at night, good sleep at night.

Fifth day - She feels well and in full power. Three dosages had been enough to cure a difficult illness quickly and thoroughly.

2.   ZOSTER (shingles)

An old lady had a shingle on one side of the lower body and experienced the usually bothersome problems. On the third day, the blisters had a blackish bottom.

Fourth day - High fever, weak feeling, benumbed of the head, the blisters turned blackish in several spots like gangrene. *Antipyreton* 3 drops. In the afternoon much heat and sweat with bad headaches in the forehead, the rash became more painful, a pulse of 108. In the evening decreasing discomfort, comfortable resting, a pulse of 88. At night, alternate sleeping and sweating.

Fifth day - The rash is bright red, but some blishters are still blackish. After the *Antipyreton* soon sweat appears on the forehead, then a big heat over all the body, and a strong pulse of 108, followed by heavy sweating without thirst; all well in the night, constant sweating.

Sixth day - In the morning still sweating, the blisters contain pus. Soon after the *Antipyreton,* stronger sweat again, decreasing in the afternoon. The urine has a thick reddish deposit, a pulse of

100. Otherwise she is quite well and hardly feels sick. During the night she partially slept and constantly sweated.

Seventh day - A pulse of 100, slight irritation. *Antipyreton,* followed by slight sweating and sedimental urine. The condition showed a change, however. She was very weak, slept a lot, had at times stinging in the side, and at night dry heat with thirst and no sleep.

Eighth day - Increased weakness, a pulse of 108, dry tongue. Only now was I told that the patient had gotten up during the previous night while she was soaking wet from sweat. She had opened the house to arriving relatives and caught a cold in the process. The illness changed now into a severe nerve fever which ended in a huge Decubitus, but she was lucky to survive.

3. FEBRIS GASTRICA NERVOSA

An old wash lady fell ill with chills, headache and vomiting.

Third day - Milder chills, diarrhea, hallucinations.

Fourth day — Heat, weak feeling, bitter taste. *Serum lactis.* In the afternoon and later again hallucinations, a restless night.

Fifth day — Mushy stool, taste less bitter. *Antipyreton* 2 drops; soon after that strong pinching in the body until the early evening without stool, fewer hallucinations. In the evening again 2 drops *Antipyreton;* again pinches in the body without stool, but not for as long, then 3 hours of sleep with heavy sweating, then sleep again.

Sixth day — A pulse of 108. Again *Antipyreton.* In the afternoon she got up and worked at washing until she was exhausted, but a relatively good sleep at night, no more hallucinations.

Seventh day - Everything is better. *Antipyreton* 1 drop.

Eighth day - She feels well again and strong.

4. SUPPRESSED HYPERHYDROSIS OF THE FEET

A young girl had been sickly for 6 weeks due to the disappearance of her extremely heavy hyperhydrosis of her feet, and she started having fever with sour belching and pain in her body. *Magnesia usta.*

Second day - The gastric symptoms disappeared but she experienced heavy pain in the whole chest area. In the evening she took 3 drops of *Antipyreton,* and soon afterwards she vomited everything.

Third day - Sour belching. The stomach had not accepted the *Antipyreton* because it had not yet been pure. Now an anti-gastric treatment was applied and on the 9th day when things were better, but the chest pains persisted, 3 drops of *Antipyreton* were administered in the evening. After that she felt a knocking in her legs which lasted for half an hour, followed by sleep, but no sweat.

Tenth day - Still chest pains, in the evening *Antipyreton,* after that again ½ hour of knocking in her legs, then sleep, but no sweat.

Eleventh day - No more chest pains, but pain in the left side. In the evening *Antipyreton,* after that one hour of knocking in her legs, then restless sleep because of the nagging pain in the side. Since the *Antipyreton* did not achieve anything, other medications were administered which finally made the hyperhydrosis of the feet reappear, thus restoring her health.

5. RHEUMATISMUS ACUTUS

A six-year old boy suffered from fever and on the third day he felt pain in the knee.

Fourth day - The right knee was swollen, the left one had pain, and pains in the left side. The pulse was feverish. The pain was so strong that the boy cried loudly and screamed, causing the parents to be quite desperate.

R.   *Antipyret. gtt. tres.*
     Aq. destill. Unc. unam. cum dimid.
     Syr. Sachar. Drachm. unam. cum dimid.
     Ms. every three hours 1 child's spoonful.

After that a much better night, but no sweat.

Fifth day - The pains spread to the feet and the hand. Lower fever, mildly moist skin. Again medication.

Sixth day - Everything was better.

Seventh day - Everything is good.

6.   ISCHLAS (Sciatica)

A lady was suffering from *Ischias nerrosa* for 8 days, which had increased each day and prevented her from sleeping. She had some fever, little appetite, more thirst.

R.   *Antipyret. gtt. sex.*
Muci 1. Gummi. arab.
Syr. Sachar. aa Drachm. unam.
Ms. half in the morning and half in the evenings.

Following the first intake there was an improvement already in the afternoon, and after the 2nd intake, she slept almost the whole night.

Second day - Mild pain, only a humming feeling in the legs, but she would not step on them. In the evening one-half of the medicine.

Third day - Improvement.

Sixth day - Improvement each day; she was able to walk a few steps. My departure to Bad Teplitz interrupted the treatment.

7.   RHEUMATISMUS DORSALIS

A lady had caught a cold by getting up at night to help her sick husband, and she experienced very strong pain in the lower back which also extended into the chest area.

R.   *Antipyreton gtt. IV.*

   Aq. Nuc. rom. Unc. dimid.

   Syr. comm. Drachm. unam.

   Ms. every three hours 20 drops.

During the following days the pain was less and on the third day it had disappeared.

8. SEDATIVUM

An hysterical lady took a dosage of *Antipyreton* at night; after that she slept more, had less sweat, much urine. On the second evening again *Antipyreton;* after that even better sleep, less sweat, much urine, but calmed nerves; she thought the drops contained opium.

A mentally retarded man suffering from hallucinations and general seizures slept calmly for 6 hours after taking a dosage of *Antipyreton,* and the following day he did not speak confusedly anymore.

9. EXAESTUATIO SAUGUIUIS

An old, but still energetic, slightly heavy lady, who had been suffering for years from a constant hissing, screaming, singing, and buzzing in her head, not her ears, experienced an accelerated tense pulse with stronger buzzing; she received 3 drops of *Antipyreton* in *the* evening, and after that she had a much worse night with increased screaming, which became even worse the next day.

Obviously the medicine had a negative effect. When *Cremor Tartan* was used the blood rushing calmed down.

This surprising positive and fast healing effect of the *Antipyreton* which I used al.so a lot without taking notes, made me value it highly and confirmed my trust in it.

Its preparation Poterius had taken from *Quercetanus* who gives it in the *Pharmacopoea,* page 675, under the title *Antidotus Lysipyretos Antimonui.* It says:

R.   *Florum rubrorum Antimonii Unc. Iv.*

Florum sulfuris sublimatorum ad perfectam albedinem Unc. II. Misceantur cum duplo colchotaris Vitrioli hungarici aut cyprii ac ter sublimentur, habebis fiores rubicundissimos, si bene operatus fusris. Ili cum aciditate vitrioli Veneris primo, deinde cum vero Saturni aethereo spiritu essentificantur.

Ilujus philosophicae essentificationis si callueris artem, medicinam, suscipicudam concinabis, veram Antidotum Lysipyre ton, omnis generis febras, pesti-feras etiam sedantem et extinquentern, si ad IV. aut V. gutlas do ea ox idoneo liquore propinareris.

The description states therefore:

1. *Floris rubri Antimonii, floris sulfuris* and *Colchotar* are sublimated three times.
2. The sublimate is first treated with *Aciditas vitrioli Veneris,* i.e. with *Acetum Aeruginis,* acetic acid from verdigris; then
3. It was made an essence with *Spiritus aethereas Saturni,* i.e. with acetone.

Lemery prepared the *Flores rubri Antimonii* by sublimating 2 parts *Antimonii crudi* and one part ammonium. The sublimate is washed out; however,

144

there is still retained one part ammonium. The
Flores has a much nicer and higher color after the
cleansing.

I have asked the chemists repeatedly which
preparation can be obtained by the sublimation, but
none could give a decisive answer without analytical
tests.

In April 1860 the pharmacist Dr. Kayser tested
Antipyreton at my request. With hydrogen sulfide a
heavy black deposit develops which turned out to be
ferric oxide when treated with hydrochloric acid.

In May 1862 the same person took up the testing
again in my presence and the results were as
follows:

1. Ammonium sulfide - heavy black deposit.
2. Potassium ferrocyanide - greenish
    discoloration; adding of potassium
    ferriccyanide brought no changes.
3. Potassium ferriccyanide - immediately a blue
    color - when heated the blue flaked and the
    liquid turned grass-green.
4. The Berlin-blue obtained with the potassium-
    cyanide turned to grass - green when
    potassium ferrocyanide was added.

5. The black deposit obtained through the ammonium-sulfide is dissolved in hydrochloric acid; nitric acid is added to oxidate the ferrum, and the result is a yellowish liquid. When *Kahi causticum* is added to segregate the oxide, a reddish-yellow deposit of ferric oxide hydrate resulted. This deposit is filtered, cleansed, and added to hydrochloric acid. When potassium ferrocyanide is added, the Berlin - blue results.

6. *Antipyreton* is added to *Argentum nitricwn,* a flaky, cheesy deposit which dissolves immediately in ammonia - a sign of chloride.

7. Added to barium chloride- white deposit which does not dissolve in much water or nitric acid - a sign of sulfuric acid.

8. Added to hydrochloric acid and hydrogen sulfide added: no change - therefore no antimony.

9. When added to potassium hydroxide and when a glass stick moistened with hydrochloric acid is held above it - white fumes, a sign of ammonium.

Dr. Graeger also investigated by titration 5 drops *Antipyreton* and determined a content of iron of 1 and 7/10 grains metal = 1,$_{87}$ oxydule = 2, 1/10 oxide in the ferrum.

The *Antipyreton* also contains (besides acetic acid and acetone)

1.   Iron,
2.   Chloride,
3.   Sulfuric acid,
4.   Ammonium.

It has a yellow—brownish color, smells like acetic acid with a faintly sweet after-taste, and a bitter, slightly sour and hot after—taste.

From the medical aspects it is highly desirable that this strong medicine may be tested thoroughly for its chemical constitution by the able hand and the keen eye of the chemist.

I made tests with mixtures of acetone and acids. Four flasks were filled with ½ drachm of acetone and respectively 5 drops sulfuric acid, hydrochloric acid, nitric acid, and acetic acid. The one with the sulfuric acid turned reddish—brown and was brownish red after 8 hours. The one with hydrochloric acid showed a faint trace of red after 8 hours.

On the second day the one with the sulfuric acid was dark, brownish-red; the one with the hydrochloric acid was slightly red; the others showed no change.

On the 6th day — Sulfuric acid dark brownish-red,
Hydrochloric acid yellowish, Nitric acid greenish-yellow, Acetic acid - no change.

On the 11th day — Sulfuric acid blackish-red, Hydrochloric acid brownish-yellow, Nitric acid greenish—yellow tinge,

Acetic acid - no change.

On the 14th day — Sulfuric acid and hydrochloric acid like before, without deposit, Nitric acid faintly green—yellow tinge, with little white slimy deposit,

Acetic acid — no change with little white slimy deposit.

After 6 weeks — Sulfuric acid black-red, no deposit, Hydrochloric acid reddish-brown like Madeira wine, no deposit.

Nitric acid faintly greenish-yellow, with little white slimy deposit, Acetic acid - no change, white with little white slimy deposit.

The cork in the sulfuric acid had shrunk considerably and was black; the one in the hydrochloric acid was less shrunk and brown; the one in the nitric acid had shrunk less and had not changed color; the same applies for the acetic acid.

Smell and Taste:

Sulfuric acid, smells like acetone, sour taste, bitter after-taste.

Hydrochloric acid, smells like acetone, sour taste, bitter after-taste.

Nitric acid, less smell like acetone, slightly etherous, no sour taste, but bitter.

Acetic acid, slightly etherous, sour taste, then burning.

For these experiments 5 drops of each mixture were added to one drachm of water.

Sulfuric acid. The drops first form a layer on the top; when the liquid is transferred it is turned brownish. After 8 hours, the mixture had the color of water, slightly turbid, and on the walls above the mixture were oil—like deposits.

Acetone smell, strong sour taste, then bitter and burning, lasting in the mouth, but no sensation izi the throat. The bitter and burning, after-taste was still noticeable the next day.

Hydrochloric acid. Mild smell of acetone, unnoticeably etherous. Sour taste, later bitter, mildly burning, also slightly contracting the mouth and sometimes belch-causing; later sour taste in the mouth, feeling of warmth near the heart, finally long-lasting bitter taste.

Nitric acid. Weak acetone smell, slightly etherous, mildly sour taste, then bitter and mildly burning, but only in the mouth, not the throat, later the feeling of warmth in the stomach. Acetic acid. The smell is hardly sour, the taste is at first weakly sour, then mildly bitter and burning, lasting for some time in the throat.

Finis.

Philosophia Maturata:

Of the Stone of the Philosophers

by St. Dunstan

Philosophia Maturata

Of the Stone of the Philosophers

by

St. Dunstan

An Exact Piece of Philosophy Containing the Practick and Operative Part
Thereof in Gaining the Philosophers' Stone

With the Ways and How to Make the Mineral Stone, and the Calcination of Metals

Published by: Lancelot Colson, Driston,
Phys. And Chym.
London, Printed for G. Sawbridge, And are to be sold at his house, Upon Clerken-well-Green, 1668.

An Exact Piece of Philosophy

Touching the Stone of the Philosophers

It is chiefly to be understood that the Ancient Philosophers did often endeavor to compose in a most short time above the Earth, those things, which by Nature, over many years were perfected under the Earth; viz., To make most perfect and most precious Sol and Luna; wherein they imitated the steps of Nature, choosing to themselves most pure Earths; white and Red, which they named their Sol and Luna; joining them together as Nature does, without repugnance, until at length they were brought to a fixation and subtlety. You must also perform this thing if you desire to obtain the desired end in this Science.

For Sol and Luna are nothing else but Red and White Earth, to which Nature has joined Argent Vive, pure, subtle, white, and Red, and so of them has produced Sol and Luna.

It is therefore needful for thee, seeking this Science, That first you get these Earths, White and Red, subtle, pure and fixed, and in these two Earths to fix two Mercuries, white in the white, and red in the red, without division, and by their least parts, so as they may endure the greatest Examen[6] of the fire, and may have such fusion, that as we see a great quantity of Water colored with a little Saffron, so they may in the least quantity abundantly tinge every metal, and all metalline spirits whatsoever, so as they be of the same Kind and Nature, and may altogether and fully bring them to their own quality.

And moreover, that in themselves they may be infinitely multiplied, and able to free the body if Man from the worst and most deadly Diseases; which Properties truly are not found in common Sol and Luna, without great Labor, (and yet but only in part) because that the Vegetative power, the Mother of all increase, for the most part, is long since extinct in them.

If you know how to perform this, and to imitate the condition of the inferior nature in making Metals, you may worthily rejoice in the name of a Philosopher, as being not meanly expert in natural things.

---

[6] high heat — PNW

It is to be noted, That the more ancient
Philosophers used not common Sol and Luna in this
Work, and therefore they said, That their work
needed not great Cost and Charges, but that it might
be as well performed by the poor, as the Rich: Which
were altogether different from the truth, if it
could not be performed without common Sol and Luna:
for they are very precious and rare, and hardly to
be gotten of poor men without great labor. Indeed,
many have bought great quantities of Sol and Luna to
nothing by this Art, and have unprofitably spent and
wasted their Time and Labor, to the destruction both
of their Bodies and Souls, which is much to be
lamented.

Moreover, in these our Times, we know no man
who does diligently and truly find out the
Philosophers Tinctures, but most of them labor
absurdly and vainly in vulgar Mercury and in common
Sol and Luna; therefore few of them obtain this
grace.

Let us take heed; for although Sol and Luna may
be subtiliated and mixed with tinctures, and so
reduced into lesser tinctures, and Elixirs with mean
profit; yet the true way according to the Doctrine
of Philosophers, is not in them: for Sol and Luna
are two tinctures Principal, red and white, buried

in one and the same body, which by nature were never brought to perfect complement, yet they are separable from their dirty and accidental dross, and afterward according to their proper qualities, are made most fit ferments for pure earth, white and red, so as in no sort they are said to breed any other thing.

For the whole Work is one, and the thing itself is one, and all the whole is derived from an Image. For our Ancestors knew, that the parts of this Stone are celestial and concrete; which were altogether absurd, if common Sol and Luna were needful to the composition thereof.

For it is said, Take a body wherein is Argent Vive, pure, clean, unspotted, and incomplete of Nature: such a body after its complete and perfect cleansing, is much better than the Bodies of Mineral Sol and Luna.

Of this same body, which is the matter of the Stone, three things are chiefly said; that it is a green Lyon, a stinking Gum, and a white Fume. But this is spoken of Philosophers, purposely to deceive Folks, and to bring them into doubts, by the many different names.

But understand you shall, one thing always is really signified, though accidentally and by names it is said to be three: for the Green Lyon, Stinking Gum, and White Fume, are spoken of one and the same subject, wherein they altogether lie hidden until by Art they are made manifest.

By the Green Lion, all Philosophers mean Green Sol multipliable and spermatick, which is as yet incomplete by Nature, having power to reduce Bodies to the first matter, and to make fixed things spiritual and flying, and so it is fitly called a Lion.

Since every Beast is subject to the Lion, so every Metalline body is confirmed and strengthened by the power of this Liony and green Sol; namely, of our Mercury, when it is Philosophically prepared. This is bred and born with a certain water, which we call Argent Vive of the Philosophers, and white Mercury. Therefore their water White and Red, gives us two tinctures, white and red, proceeding from one body and substance: There are always named our Mercuries; and after due conjunction, decoction, and digestion, we call our White and Red Stones.

By the Stinking Gum, we mean a certain stinking smell, proceeding from the unclean Body in the first distillation, which is altogether like unto stinking Asafoetida, that with a certain sweetness, whereof it is said before its preparation its smell is grievous; which is most certain. But after that, if you shall find a substance endued with these three qualities, know that it is the true matter of the Philosophers' Stone.

There rises a Question very difficult, which much troubled fantastic Heads, viz. Our Stone shows itself in a foul shape, because it is in everything, and in every place; whence many men reading this, make choice of several and stinking things, which with great labor they distil, calcine, and join together. But let such hear what the Philosophers say, "Who so seeks the Philosophers Secrets in Turds loses his Labor, and in the end finds nothing but deceit."

Yet there is also another thing which troubles these men's Brains, viz., Our Stone is bred between two mountains, it is cast out into the Dunghill, and trodden under men's feet, it is counted a most vile and contemptible thing, it is generated between Male and Female, and lies hidden in You, in Me, and in such like things. And contrarily it is said, Our Stone cannot be in things differing from its kind, namely, from the Nature of SOL and LUNA; for it can give nothing that it does not have. A Nettle cannot produce a Rose, nor a Woman a Dog; how then shall we resolve so many doubts rising from Contraries? Truly, it is easily done for it is plain that nothing in this World, whether it is Animal, Vegetable, or Mineral, can be generated without a natural and a special appetite.

Therefore according to the Doctrine of Philosophers, which informs us only by obscure Examples, we must understand that the Stone may be by Similitude in everything, and in all places, chiefly, because it is nothing else but a specific virtue and quality joined with natural heat, whereby every compounded thing is brought to his most perfect determined end.

Things generally spoken are always generally to be understood, for what earthly thing can be in very thing, and in all places, but only a specific Appetite, and a natural Heat; for these are the immediate and near causes without which the Stone cannot be.

Whosoever therefore desires to understand the Stone, let him not depart from his specific quality and Original.

Of a Man cometh a Man, of a Rose a Rose; so likewise from a matter which is potentially Gold, having things necessary, and Excrements purged, arises SOL by an inward Appetite; therefore from a Metal arises a multitude of metalline tinctures and perfection. The Stone is made of a metal, living, hot and moist, when natural heat is joined with it, whereby it is made apt to generate its like. For our Stone is most pure matter, viz. the nature of Sol containing in itself a vegetable heat, whereby it has power and virtue always to multiply in itself a vegetable heat, whereby it has power and virtue always to multiply in its own specific and natural form; therefore it is called the secret Fire of Nature, stirring up the compound, and perfecting it in our Glass into a Stone, in like manner as seeds by reason of its own proper natural heat, and radical moisture, if its mother Earth does putrefy to admirable Generation and Multiplication. Whosoever therefore keeps not this our heat, our fire, our Balneum, our invisible and most temperate flame, and of one regimen, and continually burning in one quality and measure within our Glass; I say, whosoever understands not this Dunghill, horse belly, and moist fire, shall labor in vain, and shall never attain this Science.

You see therefore that the radical humidity, which is that first vegetable Virtue, is the cause of multiplication of everything in its kind. Therefore of the Composition of Sol and Luna take our burning water, that Aqua Vitae, which the ignorant do think, but falsely, to be extracted from Wine, Oil, and such like Liquors.

I say, such green Sol and Luna, in which the vegetable virtue is not extinguished, but is living, hot and moist, and has power to reduce all Bodies to their vegetability; for by this, with God's permission, bodies extinct, and not multiplicable, may more easily get the habit and virtue to germinate, which of Philosophers is called the beginning and term from whence the Stone is generated.

Marie the Prophetess in an Epistle to Aron, writes, that "The Body taken from little Mountains, is a Body white and clear, not suffering putrefaction nor motion, and it is that which is generated between Male and Female." By these little Mountains is understood Sol and Luna, which are naturally separated from us by a great distance, by whose influence Gold and Silver are generated, both which are in our Mercury.

By Male and Female, we understand Agent and Patient, Active and Passive, both which are also in our active Mercury, and in our Passive Earth. Whereby, without doubt, it is inferred that mineral Earth and Water are the Active and Passive matter, as the Philosophers' Stone. And here hence appears that community between the Poor and Rich, seeing that the Stone may be made of one thing, without visible Sol and Luna.

But here by the way I advertise to you that between the Elixir and the Stone there is this difference: for the Stone rejoices in unity and simplicity, but the Elixir in plurality.

The Stone therefore is one thing, our Mercury, Sol, Luna, our Tincture white and red, which maybe naturally joined with its own proper Earth, or with the Earth taken from the little Mountains, and may easily be obtained by mortal men. But the Elixir is the same Vegetable Mercury, which yet by reason of its fixation, is said not to be common, but consisting of many things, for it is absolutely fixed in the Earth of common Sol and Luna. And therefore it always consists of many things, viz. of Mercury vegetable, and of a different Earth, which is neither Common nor fit for poor men. But of this

Earth it is not much to be respected of what
substance it is, so it is fixed.
Alphidius is if the same Opinion, saying, "The
Faeces from whence this Earth is taken, Seeing it is
of no value, is altogether to be rejected, and the
Mercury to be planted in another Subtle Earth." For
its own Earth is seldom natural in composition of
the Elixir.

Yet, my Friend, I will name it to you by its
own Name, whereby the common People name it; and it
is the end of the Egg, whereby we understand the
nature of Metals, viz. Mercury rightly mixed by
Nature, with its own Sulphur, of its own accord
inclined by putrefaction and growing.

From this Egg three things must be considered;
namely, the Yolk, the White, and the Shells; and
this last is only and altogether necessary for
Philosophers, which is called the end of the Egg;
that is, the last part, rejoicing in perfection,
having the likeness of a little Mountain, and also
generated between Male and Female; which when it is
perfectly calcined it exceeds all Earths whatsoever
in whiteness and subtlety, enduring the greatest
fire embracing Tincture, and desiring a metal in
Nature, which is hardly believed of Workers in the
Art, unless they being overcome by experience the

Mistress of things, that they be compelled to confess and admire it. But another Earth, wherein there is any Mercurial humidity, will not drink up our Mercury with so much greediness, and therefore it is not so commodious; the reason is, because it abounded with its proper and natural humidity was naturally transferred to the generation and union of the White and the Yolk.

Yet we deny not, that the virtue is necessary, yet profitable for the preservation of man's Body, which is derived from the outward parts of the inward, yet so as it be mixed with the Elixir of life this Earth is wholly condemned, when the matter included is corrupted, and is cast out on the dunghill, and everywhere trodden underfoot, and accounted unprofitable.

Sometimes I, desiring to try whether it would join with our unctuous humidity, put it thereto, which it drunk up with so great an appetite as that it seemed spongy, and a most fat congelation, rather than an Earth naturally naked. After I had gently evaporated the Mercury, it remained very Citron. Here I will end and show what matter and what way of practice is necessary for this Work and Art.

Of the Practical and Operative Part

In the Name of God take a Drop of the Green Lyon, which I have mentioned before, and dissolve him in distilled Vinegar very well for ten days, stirring the Compound strongly three times every day, that it may be well mixed; then separate the Faeces three times by filter; afterward evaporate the Vinegar with a gentle fire, until it be thick as Pitch, then pour it out, and keep it safe.

Having 12 pounds of the Green Lyon thus brought into Gum, you may believe, that you have seen Earth of Earth, and the Brother of Earth, whereof Philosophers have often spoken; put thereof three pounds into a Glass, whose third part may contain at least four Sextaries of Wine; put it into a Furnace with Sand, so as the sand may be two fingers thick under the Glass, and about and above the Matter; then the matter being a little dried with a gentle heat, put a Receiver not yet luted thereto; after a few hours, having received a certain Light Water, when you see a certain white Fume begin to ascend, put thereto another most long and most large Receiver, which lay close, lest the Spirits break forth, which are most necessary in this Work:

Note also, that from the first appearance of the White Fume, the fire must be discreetly increased little by little: This same tinges the Receiver with a certain thick and milky humidity which is our Luna, and therewith shall also ascend a most red oil, called the Philosophers aerial Gold, a stinking Menstruum, the Philosophers Sol, our Tincture, burning Water, the Blood of the Green Lyon, our Unctuous Humidity, which is the last comfort of Man's Body in this Life; the Philosophers' Mercury, the Solutive Water, which dissolves Sol under conservation of its Species; it has also may other Names.

Continue this Distillation from the first appearance of the white Fume 12 hours following: then remove the Receiver, and stop it close, lest the Spirits be lost, which are very volatile and penetrative: And thus you have the Blood of the Green Lyon, called The Secret Water, and most sharp Vinegar, by which all Bodies may be reduced to their first Matter, and purges Man's Body from all Infirmities.

This is our Fire always equally burning in one measure within the Glass, and not without: This is our Dung-hill, our Horse-belly, working and producing many Wonders in the most secret Work of

Nature: It is also the Examiner of all Bodies
dissolved, and not dissolved; a Fire hot and moist,
most sharp, a Water-carrying Fire in its Belly;
otherwise it could not have power to dissolve Bodies
into their first Matter: This is our Mercury, our
Sol, our Luna, which we use in our Secret Work.
Take the Faeces left in the bottom, as soon as they
are cold; for they are our Cross-Bill, far blacker
than Pitch, which you may set on fire by putting a
kindled Coal into it; so as they shall be calcined
of their own accord into a most Yellow Earth: But
this Calcination suffices not for its perfect
cleansing; put it therefore into a Reverberatory,
with a moderate heat, for eight days, and so many
Nights following, increasing the heat and flame,
till it be white as Snow; they may also be calcined
in a Potter's Furnace, being meanly hot.

Having this white Earth, you may putrefy and
alter it, or the Calxes of other Metals prepared; as
I will teach you in that which follows, at your
pleasure, into a new whiteness or Redness, by means
of our Luna or Mercury, which putrefy with them by
Generation, and Vegetation; which Properties they
wanted before: for the Philosophers say, first
calcine, then putrefy and dissolve; distill,
sublime, descent, and fix often with our Aqua Vitae;
wash and dry; and make a Marriage between the Body

and the Spirit; and if the Water be congealed, by a natural commixture with the Body, then the Body shall die of the Flux, shedding its blood, and putting on many Colors; after the third Day, he shall ascend and descend, first to the Moon, then to the Sun, through the round Ocean Sea, and without end, sitting in a very little Ship; and when his Journey is ended, he shall not need any great expense. And thus you may wait again patiently for the Harvest, and you should be filled with Joy and Riches. And now we will speak of Putrefaction. Take an ounce of this Calx hidden before in the Philosophers Egg, and thereon put of the red tincture to cover it two fingers; then seal it, and set it to putrefy eight days in a most cold place; which being ended, it will drink up the humidity; again pour on as much of the tincture, and let it stand as before for other eight days, continuing again the said imbibitions and times: Let it stand till it cease to drink any more tincture; remove it not from its place, until it be blacker than any pitch; which being seen, set it into a natural balmy, that the moisture with the black earth may be digested, and fixed into a white Mineral: then divide it into two equal parts, and work the one for the white, and the other for the red stone, which you can thus easily perform: Ferment the one part with the oil of Luna, that is, with white water, and the other part with the oil of Sol, that is, with the red water. By greater heat and digestion, it shall be converted into a most red powder, like Dragon's Blood.

This powder being joined with a part of our Mercury, and circulated, is called Aurum Potable, Elixir of Life and of Metals, which transmutes Mercury and all that is imperfect into most perfect Sol.

But here learn a general Rule: if you ordain the Elixir only for the white, then keep one part of the red work, distill the other part with a gentle fire, taking the white water, which we call our white tincture, our Eagle, our white Mercury, and Virgin's Milk, and having these two Mercuries, you may practice with them, either upon their own Earths, or upon the calxes of Metals prepared. For it is said, the Earth is not much to be respected, so that it be fixed. Therefore take whichever you want, being first altered into whiteness; and for the white work, you may ferment thus:

Take the Calx of Luna and earth altered, in equal parts. Grind them together, and temper them with white Mercury, named Virgin's Milk, which keep safe; sublime the rest not fixed, and that which arises to the sides of the glass, like mercury sublimed, reiterate upon his proper calxes, grinding and tempering with our Virgin's Milk, distilling and subliming as before, until no fire will raise it. This is our Mercury sublimed and fixed, made of the white Earth of bodies altered, arising at first admirably by the virtue and help of water.

This is that Mercury, instead of which the unlearned take the compounded of common Mercury, Vitriol, and Sol sublimed; wherein they are deceived. When this is thus fixed into white earth, it is afterward calcined, whereof is made an Elixir or stone as follows.

Put it into a circulatory, and pour thereon Virgin's Milk to cover it, then circulate it to the thickness of oil, by drying and calcining it, as often as you will, for by this means it may be augmented infinitely.

But before you make projection, congeal it into an oily powder, one part thereof converts a thousand, nay ten thousand parts of Argent Vive, and the other metals into pure Luna, enduring all trials.

In like manner you shall work with the Red
Water upon the calx of metals, by fermenting and
subliming upon the calx of Sol altered.

And note that you cannot have a perfect
ferment, until it is altered with Mercury from their
first qualities, into a new whiteness and redness by
means of Putrefaction and alteration, which before
it wanted.

But when after putrefaction, it shall be
reduced into Whiteness, then it becomes spiritual,
and is more apt to join better with our Mercury
sublimed naturally, and by the least parts, and also
to be fixed together inseparably; which would not be
so natural, if one part were fixed, and the other
part to be separated.

Moreover, when spirits have not virtue to
penetrate bodies, nor bodies to embrace spirits, it
is impossible that they should be joined by their
least parts.

But contrarily, when ferments are made
spiritual, then spirits will join with spirits, and
the body which was most perfectly fixed, is
naturally disposed, and inclined to return to his
former fixation; which without doubt cannot possibly

be in bodies which were never perfectly fixed, but
the body before fixation, desiring a solid habit and
fixation, draws with him, and into his disposition,
all spirits whatsoever, which are joined with him,
and not degenerating as Sulphur Vive, Arsenic
sublimed, Bole-Armonick, and such like.

Common Mercury sublimed, may very well be
joined with spiritual ferments, which with calx of
ferment not altered, will never be perfectly joined.
Therefore this part of natural Philosophy excludes
all citrinations and dealbations, which were not
produced by a perfect alteration before the
tincture, were joined to their bodies and spirits.
For nothing can be made an Elixir, until it has
passed the Philosophical Wheel; which being unknown,
all labor comes to nothing.

An Abbreviation of the Work

Wherein almost all Elixirs are Contained, and the Ways to Make Them

The First Abbreviation

Take Vitriol, calcine it into ashes, then beat them into most subtle powder; put them in an Urinal, and pour thereto Virgin's Milk to cover them, stop the Urinal, and pour thereto Virgin's Milk to cover them, stop the Urinal with a Lien cloth, and let it stand eight days, then add thereto as much of the aforesaid Milk, reiterating this from eight days to eight days, until a certain Crystalline earth like Fishes eyes, appear in the upper part thereof, which separates from the gross part remaining in the bottom, which put into the Philosophers' Egg to digest discreetly until it be perfectly fixed. Then increase the fire, till it be perfect yellow, and then again increase the fire, until it be red as Dragon's blood: Then add to this a part of red Mercury, to cover it, and congeal it by Circulation into Oil, and afterward into Powder, and do thus three times.

Project one part of this Powder upon forty of
most pure Luna, melted with one part of most fine
Sol, and it shall be converted into most pure Sol;
or if you project it upon Amalgam of Mercury and
Sol, or of Mercury and Luna, it shall be more
certain, and more plentiful.

But if you will have Gold most perfect, and
most high, take the Elixir out of the Egg, put it in
an Urinal, and pour upon of the foresaid red
Mercury, equally compounded, and mixed with a strong
Corrosive made of Vitriol and Sulphur, which
evaporate from the Elixir with a most gentle fire,
and by means of the tincture of the one water and of
the other, shall be fixed with the Elixir, by
augmenting its quantity and color; which being often
repeated, the Elixir shall be converted into the
form of Oil, in which if you quench Lamins of Luna
annealed, they shall be throughout tinged into most
perfect Sol, which being melted with a part of most
pure Gold, it shall be purer than any common Gold.
But if you take as much of the white Earth of Mars
altered as before, of Vitriol, fixing it upon the
calx of Sol altered, and afterward rubified, and
then convert it into Oil as before, with the said
compounded water; you shall have a great Elixir,
converting every Metal into most pure Gold.

This work may be done in twelve weeks; but it is not good for the health of Man's body.

In the same manner with the ferment of Luna altered, you may fix the white Earth of Vitriol, and of Mars altered, which are reduced into Oil, with the aforesaid Virgin's Milk, being equally mixed with water of common Mercury, sublimed, fixed, and calcined, so have you the best Elixir to convert all Bodies into most pure Luna.

The Second Abbreviation

If you cannot artificially prepare the aforesaid white or red water, you may far sooner attain the end of the work. First therefore fix Mercury sublimed and calcine it, and then dissolve it in the other Mercury white or red, until they are made one water: purify the water three weeks, and it will alter the calxes of any Metal; for in this work is joined a twofold Water, namely, natural and against nature.

The way to fix Mercury sublimed, is thus: First sublime Mercury; if there be half a pound of it, join thereto half a pound of Saltpeter, and as much of Vitriol, grinding and tempering the mixture with distilled Acetum, till all become like white Paste; when they are thus incorporated, sublime seven times, that of his own accord he may be clear, then fix it in this manner: put two or three pounds in a long receiver, stop the mouth, place it in ashes, so as the Globe may be very wholly covered; the first week, give it a gentle fire; the second week stronger, and the third week most strong; this done, it shall be very well fixed. Again, dissolve it in Virgin's Milk, after the aforesaid way and order; if you are in need of money, you may obtain a branch or particular in far shorter time.

Thus, take the aforesaid white compounded
Mercury, and fix it upon the calx of Luna, not
altered by circulating it thereon, and when one part
is fixed, add more, repeating it often, until the
calx itself melt like Butter on a fiery Coal.

One part thereof projected upon ten of mercury
purged, makes good Luna for Vessels and Household
Ornaments. This same way you may handle our
Mercuries composition, being made as before with the
said water extracted from Mercury sublimed, fixed,
and calcined, and dissolved in the said red Water,
so as it be then calcined upon the calx of Sol not
altered, and you have the best Tincture to convert
Sol and Luna, whereof Rings and other things may be
made.

The Third Abbreviation

Put into a Circulatory an ounce of the calx of the Egg-shells very well reverberated, and pour thereon of white or red Mercury to cover it; then nip the glass, or stop it close with lute made of powder of Iron, Vitriol, and Honey, well boiled together, circulating in Balneo, till it be dried up into powder.

This done, pour in more observing the same order until it be made oil; this converts Mercury, and the other Metals into most perfect Sol and Luna, according to the nature and disposition of the Elixir.

After the same manner you may circulate our Mercury upon the Calxes of Metals. There can be no way shorter than this; for if you put an ounce of calx of Sol with mercury, before likewise fixed, and pour thereto as much red Mercury as may cover the calx two fingers breadth, then stop it close with paste compounded with Honey, Bole-armenick, and Iron dust, mixed and strongly tempered and boiled, till it be stiff and black.

Then set the circulatory in a Furnace, and with gentle heat digest the red Mercury into a red and

fixed calx, then add thereto as much more of that
Mercury, circulate and dry it as before, till the
Calx have drunk as much Mercury as it can, and be
converted into a thick blackish oil, and so you have
an Elixir which converts ten parts of Mercury,
purged and heated, into a most red powder dry and
fixed, which if you also put into a Circulatory with
increase, and digest by imbibition and congelation,
as before, it shall be so much increased in
quantity: And thus you may multiply this Elixir
infinitely.

One ounce thereof will congeal a hundred of
crude Mercury into powder; one ounce will convert
ten of any metal into most pure Sol.

And this way you may work with the Calx of Luna
and Mercury joined together, so as evaporation be
made by Circulation, and adding our natural white
Mercury, until it be reduced into oil, proceeding in
all points as in the former, with the red mercury
upon the calx of Sol, and so you shall have a white
Elixir converting all bodies into most pure and
perfect Luna.

The Fourth Abbreviation

Take an ounce of the Earth of the Quintessence, smelling most sweetly, and an ounce of the Mercury of Virgin's Milk; powder the Earth, and join it with the mercury.

This way shall be made a perfect composition in the first order for the white Elixir, which by longer time and greater fire is reduced into a red Elixir; put therefore the compound into a blind Urinal (as it is called) very close stopped, and digest it in dung equally for fifteen days; then take it out, and shut it up in a Philosopher's Egg; and digest it in a gentle heat, till it be black, and so into perfect whiteness. This we call a white Elixir: within this time, the fire being increased, will be red, of which one ounce cements hundreds of ounces of Mercury into Sol.

To multiply it, take a part thereof and join it in the foresaid manner with Virgin's Milk, digest as before, unto whiteness, and then unto redness. In this second repetition, the Projection will be upon four hundred. By this Projection you may multiply it at your pleasure.

The Fifth Abbreviation

Dissolve the red Calx of Sol and Mercurie in the first most strong Corrosive composed of Salt-Peter and Vitriol the common way; put the solution in a pelican in Balneo; drawing off the one half, then stop it most close, dry it up with a gentle heat; then add more of the Corrosive, observing the foresaid order, in dissolving, evaporating and congealing ten times, until the Corrosive cease to arise; which is then done, when by no fire it can be fixed into powder, but remains like oil and thick. This Elixir converts Mercury, and every metal into most perfect Sol; This work ought to be done in a Circulatory placed in an Earthen Pot, wherein it must stand covered with dung to the middle; This Pot must be full of holes in the bottom, and must be placed upon the mouth of a Copper Vessel half filled with hot water, as a Copper Vessel is placed of a Furnace, wherein fire must be made to be continued discreetly for the necessity of digestion. This Experiment is called Rustum.

Of the Mineral Stone

God is wonderful in his Works, who is Virtue, teaching Truth.

Take in this name of Mercury white or red, simple or compounded, and dissolve therein five stone of the Sea; Doing in all things as you did in Vitriol, and you shall have the great Elixir.

By the same way of putrefaction, all Minerals may be altered, and so of every fixed thing (a due matter being added) may be made an Elixir, for our Mercury white and red must be joined with fixed things which want Mercury; and this way the Metalline Bodies may be brought into a metalline form, namely a Vitrified Powder, as also Egg-shells, which when they are perfectly calcined, will endure fire more than Sol; and thereof being well and artificially tinged, Philosophers have made Sol in the space of one day, which nature cannot do underground in a thousand years; A thing hard and incredible to the unlearned; yet true and most certain, and confirmed by the testimony of many men. Be therefore not solicitous or curious in choosing your Earth so that it be of a Metalline Nature, and enduring the Fire.

Hereby Glass is made malleable, and by means of this Tincture, is converted into transparent and fixed Metal, whereby it appears that this science is possible.

For there is no Earth which does more easily embrace the Spirituality of our Mercury, than that which is most deprived of Mercury, and moisture; which Privation you shall not find in Bodies of another Nature, although yet they be very much calcined.

Wherefore it appeared manifestly, that seeing Sol and Luna, are nothing else but Earth, Red and White, wherein a most pure Mercury is fixed and joined by the least parts, that Philosophers (having the same elements) may artificially imitate Nature in her Composition under the Earth, to produce the same effect: for it is certain, that earth may be fermented to Water, so as it be fixed; and water fermented to Earth, if it be perfect and cleansed; and this without the help of any common Sol or Luna: And therefore Philosophers in their Writings have taught, that the Stone is equally common to the Poor and Rich.

These things considered, you shall understand, that our Stone lies hidden, and secretly lurking, often in places least expected, and nothing esteemed, whose matter and nearness, if it should be known, would produce most great danger.

It is to be noted that the Philosophers have found out diverse ways of handling this one thing: but I answer for them all, and briefly conclude, That our Earth does drink up and fix our Mercury; and that this mercury washes and tinge our Earth, and thus to perfect it into the Stone, without any further ferment.

For the white Mercury gives a most perfect Tincture of Luna, and the Red Mercury of Sol: Therefore, when they are fixed in convenient Earths, they make Sol and Luna, without any help of common Gold and Silver. Behold that you understand this Tincture, which we draw out from a vile thing of no price: yet note, that he that has Salt in his Breast, may ferment this Tincture with common Gold, whereby he may obtain incomparable Riches, yet with Wisdom, with most great Cost, and not without danger.

For from Sol alone, by means of this Tincture, which is our burning Wine, is made a most precious and a most perfect Elixir, white and red; for it rejoices in fullness of white and red Sulphur, whereby may be made most perfect Silver.

Of this Work, I have written more fully in my Seventh Book; wherein I treat of the manifold plenty of Gold, and of the greatest Elixir of Life: But here also I will briefly touch it. Understand therefore, that it behooves you to alter the Calx of Gold (with the aforesaid Stone, equally mixed with the Water of mercury, sublimed and perfectly fixed) into most white and fixed Sulphur.

Then calcine it well, that the strength and poison of the Fire against Nature put to it, do hasten to Putrefaction and alteration, may be utterly destroyed.

Then imbibe it with the foresaid simple Milk, until the Calx itself shall have drunk up a reasonable quantity thereof, and that it be fixed. Dissolve it again with the same Milk, and make it volatile; afterwards fix and calcine, and then bring it into Oil, with a little part of that Virgin's Milk by circulation, and so it shall be a perfect Elixir, converting Mercury, and each imperfect Metal, into most perfect Luna: and by the same way, you may rubify the other part without Red mercury, by fixing and calcining, and afterward dissolving it with the same Red Menstruum, and at last by circulating it into a thick Oil, which we call potable Gold, a curing and preserving Elixir of Life, and of Metals.

Know also, that if our Red Mercury equally with mercury sublimed and fixed, be circulated with Lute, Vitriol, or Iron, before and after Rubification, be digested into Oil, it will convert thin Laminae of Luna annealed and injected into pure Sol, which if you afterward take it out, it will serve for all need to live withal.

It is a general Rule, That if you will be a Master of this Art, it is needful to make all Medicines gummous (gum-like, or composed of gum. – PNW) and fusible, melting like wax of their own accord, without Fume, upon a plate annealed.

For by this means, each part will follow the other in Projection and will jointly dilate themselves through the Pores of the Metal, without any disjunction: but if any part be ponderous, it will separate the parts of the Metal, and make it brittle.

Therefore the Medicine must often be subtiliated, after that it is perfectly fixed, that

186

at least it may be an incombustible Oil, and rather
may be called a Species, then a Genus, because it is
nothing else but a fixed Tincture of Color.
If (this thing observed) you cannot prepare thy
Medicine thus, you shalt make fair Metals, or else
not.

Also here understand, two Bodies to be
dissolved with the Natural Menstruum, is always the
second Calx, not the first; and therefore it
behooves you to dissolve Calx of Metals with a
compound Mercury, as before is taught, that they may
sooner putrefy, and be altered into the second Calx
(which we call Sulphur of Nature, and Foliated
Earth) which we then dissolve and coagulate
(circulate) into Oil, with a Simple Menstruum;
namely, Natural.

The Calcination of Metals

Now learn how Metals are to be calcined: Know therefore, that Saturn and Jupiter we calcine only one way, which is this: Put either of them into a great Iron Vessel, and in the Fire, so that the Flame may beat upon the Metal; and draw off the Scum with an Iron Rake, to the sides of the Vessel, stirring it often, until it grow white then cease it, and gather the subtle Powder; one Ounce is sufficient for thee.

Sprinkle Venus and Mars with the best Vinegar well distilled, that they may gather Rust: burn this with most strong Fire in an Iron Dish, when it is red-hot, cool it in the best Acetum, evaporate that Acetum, and gather a most red Earth; which dry, and keep safely.

Amalgam Sol and Luna, and grind it on a Marble with Powder of Salt, prepared without any moisture, until no Mercury appear: then sublime and evaporate the Mercury with strong fire; grind that in the bottom, into most subtle Powder, and sublime, until no Mercury remain with it; wash the Calx with hot Water, to take away the Salt; dry it, and you shall have a Calx more subtle than Meal.

Another way is thus: Take thin Lamins of Sol, anneal and cast them into Mercury, heated on strong Ashes; so the Mercury will drink up the Sol.

Note, That every Ounce of Sol requires four and twenty Ounces of Mercury; put this amalgam in a large Glass; bury it in Sand in a great Furnace, give it Fire by degrees; after the sixth hour make it vehement: continue this heat five days and nights, at each hour putting down the Mercury which ascends, with a linen Cloth, bound with a little Iron Rod, and stopping the Glass with Lute, till at last all become a Powder redder than Blood, which then we call the first Calx, good and perfect; with which, if you mix the Fire of Nature, to use his Virtues, as it requires, though one cannot err in this Science.

The Recapitulation

I have told out of what, and how you shall make our Mercury white and red, and how this Mercury is to be actuated and sharpened; how you shalt prepare Calx, how to purify and alter them into a new Whiteness, which we call our Mercury sublimed; how to abbreviate the time of Putrefaction and Alteration; how to fix and dissolve again, and then how to circulate into a white and red Elixir; how by imbibitions, with proper Waters white and red, they may be infinitely multiplied to an incredible profit.

Learn therefore Patience, fear God and love Him, keeping these Secrets, and then the Lord will bless your Endeavors.

Finis.

A Word from the Publisher

Thank you for purchasing this small work from The R.A.M.S. Library of Alchemy. During his lifetime, Hans Nintzel was dedicated to the identification, acquisition, study, retyping and, when necessary, translation of what he considered to be the most important known works on Alchemy. Hans was assisted by his sparse network of fellow Alchemists, all members of the Restorers of Alchemical Manuscripts Society (R.A.M.S.). I was an active member of R.A.M.S.

My goal is to publish all of the works originally made available through R.A.M.S. as photocopies. To facilitate this, I have chosen to have the books professionally printed. I also have a few titles that I intend to add to the original R.A.M.S. Library, selected by strict criteria established by Hans.

If you have a work on Alchemy that you believe should be a part of the R.A.M.S. Library, please contact me through R.A.M.S. Publishing Company.

Philip N. Wheeler

www.ingramcontent.com/pod-product-compliance
Lightning Source LLC
Chambersburg PA
CBHW080807180526
45168CB00006B/2357